Communities and the Clean Energy Revolution

Environmental Communication and Nature: Conflict and Ecoculture in the Anthropocene

Series Editor: C. Vail Fletcher, University of Portland

This interdisciplinary book series seeks original proposals that examine environmental communication scholarship. In the Anthropocene era, the period during which human activity has become the dominant influence on climate and the environment, the need for highlighting and re-centering nature in our worldviews and policies is urgent as collapsing ecosystems across the globe struggle to survive. Topics might include climate change, land-use conflict, water rights, natural disasters, non-human animals, the culture of nature, ecotourism, wildlife management, human/nature relationships, food studies, sustainability, eco-pedagogy, mediated nature, eco-terrorism, environmental education, ecofeminism, international development and environmental conflict. Ultimately, scholarship that addresses the general overarching question, "how do individuals and societies make sense of and act against/within/out of nature?" is welcomed. This series is open to contributions from authors in environmental communication, environmental studies, media studies, rhetoric, political science, critical geography, critical/cultural studies, and other related fields. We also seek diverse and creative epistemological and methodological framings that might include ethnography, content analysis, narrative and/or rhetorical analysis, participant observation, and community-based participatory research, among others. Successful proposals will be accessible to a multidisciplinary audience.

Recent Titles in This Series

Communities and the Clean Energy Revolution: Public Health, Economics, Design, and Transformation
By Melanie J. La Rosa
Fracking and the Rhetoric of Place
By Justin Mando
Water, Rhetoric, and Social Justice: A Critical Confluence
Edited by Casey R. Schmitt, Theresa R. Castor, and Christopher S. Thomas
Environmental Activism, Social Media, and Protest in China: Becoming Activists over Wild Public Networks
By Elizabeth Brunner
Natural Disasters and Risk Communication: Implications of the Cascadia Subduction Zone Megaquake
Edited by C. Vail Fletcher and Jennette Lovejoy
Critical Environmental Communication: How Does Critique Respond to the Urgency of Environmental Crisis?
By Murdoch Stephens
Natural Disasters and Risk Communication: Implications of the Cascadia Subduction Zone Megaquake
Edited by C. Vail Fletcher and Jennette Lovejoy
Communicating in the Anthropocene: Intimate Relations
Edited by C. Vail Fletcher and Alexa Dare
Communicating the Climate Crisis: New Directions for Facing What Lies Ahead
By Julia B. Corbett

Communities and the Clean Energy Revolution

Public Health, Economics, Design, and Transformation

Melanie J. La Rosa

LEXINGTON BOOKS
Lanham • Boulder • New York • London

Published by Lexington Books
An imprint of The Rowman & Littlefield Publishing Group, Inc.
4501 Forbes Boulevard, Suite 200, Lanham, Maryland 20706
www.rowman.com

86-90 Paul Street, London EC2A 4NE

Copyright © 2022 by The Rowman & Littlefield Publishing Group, Inc.

All rights reserved. No part of this book may be reproduced in any form or by any electronic or mechanical means, including information storage and retrieval systems, without written permission from the publisher, except by a reviewer who may quote passages in a review.

British Library Cataloguing in Publication Information Available

Library of Congress Cataloging-in-Publication Data

Names: La Rosa, Melanie J., author.
Title: Communities and the clean energy revolution : public health, economics, design, and transformation / Melanie J. La Rosa.
Description: Lanham : Lexington Books, [2022] | Series: Environmental communication and nature: conflict and ecoculture in the anthropocene | Includes bibliographical references and index.
Identifiers: LCCN 2021048335 (print) | LCCN 2021048336 (ebook) |
  ISBN 9781793639226 (cloth) | ISBN 9781793639240 (paperback) |
  ISBN 9781793639233 (ebook)
Subjects: LCSH: Communities—United States—Case studies. | Clean energy—Technological innovations—United States—Case studies. | Renewable energy sources—United States—Case studies.
Classification: LCC HM756 .L3 2022 (print) | LCC HM756 (ebook) |
  DDC 307.0973—dc23/eng/20211117
LC record available at https://lccn.loc.gov/2021048335
LC ebook record available at https://lccn.loc.gov/2021048336

♾️™ The paper used in this publication meets the minimum requirements of American National Standard for Information Sciences—Permanence of Paper for Printed Library Materials, ANSI/NISO Z39.48-1992.

Dedication
To Sister Helen Werner,
who showed me the first solar-powered device
I saw someone use in daily life:
a passive solar oven on your patio at the Maryknoll
mission in Lemoa, Guatemala.
You were before the times!

Contents

Acknowledgments		ix
1	Introduction	1
2	The Front Lines of the Clean Energy Revolution: The Beginning	11
3	Atlantic City, NJ: Offshore Wind and Mixed Generation	23
4	New York, NY: Green New Deal	51
5	Colchester, VT: Green Mountain Power	77
6	Salinas, Puerto Rico: Solar for Survival	95
7	Isabela, Puerto Rico: Design for Survival	129
8	La Riviera, Puerto Rico: Resilient Infrastructure	149
9	Highland Park, MI: Solar-Powered Streetlights and Community Activism	159
10	Las Vegas, NV: 100% Renewable Energy for City Power	183
11	The End? or Another Beginning?	205
Additional Resources		209
Bibliography		211
Index		229
About the Author		237

Acknowledgments

The most important acknowledgment is recognizing the leadership and talent of all the people truly on the cutting-edge of the clean energy revolution. That includes people who shared their stories with me and those whose efforts in previous decades laid the groundwork for the projects profiled in these pages. All empower others to find ways of using clean energy to make their communities cleaner, more resilient and more sustainable.

I would like to thank all of the people mentioned in this book for their time and generosity in sharing their experiences. I spoke with dozens of people during this research, from phone calls during the film's development stage to answering questions on camera, allowing me to film them over many months, and providing follow-up information as I edited the film and wrote this book.

I am grateful to the interview participants, who are Nicole Acosta Martinez, Mark Baker, Nick Baudouin, Kyle Bolger, Jennifer Bolstad, Rebecca Bratspies, Janes B. Caldero, Monica Coffey, Council Member Costa Constantinides, Ali Dirul, Rick Dovey, Pat Egan, Karanja Famodou, Jeanne M. Fox, Paul Gallagher, Anthony Gigantiello, Mayor Carolyn Goodman, Robert Wylie Hyde, Jackson Koeppel, Joshua Langdon, Paul Lee, Stephanie López, Emmanuel Maldonado, Margaret (Peggy) Matta, Thomas Meyer, Walter Meyer, Gregg Murphy, Nandi , Shimekia Nichols, Efraín O'Neill-Carrillo, Tom Outerbridge. Paola Pagán Berríos, Mary Powell, Katia M. Rivera Cruz, Yanila R. Rivera, Alejandro G. Rodriguez, Lyon Roth, Shayra E. Sáez Lòpez, Juan A. Santana, Juan Santana, Ruth Santiago, Juan Shannon, Travis Tench, Maria Thomas, Robert J. Thomas Ramirez, Vidal Torres, Jariksa Valle Feliciano, Diane Van Buren, Carmen A. Velasco-Román, Marco Vellotta, and Chris Wissemann.

In addition to the interview participants portrayed in these chapters, many others contributed to this research, including Iván Alvarez, Michael J.

Alvarado, Bob Bunao, Kristin Carlson, Gustavo A. Carrión, Josh Castonguay, Rafael A. Cintrón, Erin Cole, Amy Cook-Menzel, Carmen Bras, José Matos Cruz, Daniel de Jesús, Carmen M. de Jesús Tirado, Dr. Catalina de Onis, Megan Doran, Bryan Duncan, Samy Esayag, Ian Graham, Juan R. Ibañez, Jim Leidel, Constance Lilley, Samuel Lozada Rosario, Emmanuel Mercado Pérez, Lance Miller, Dr. Dustin Mulvaney, Eadoain Quinn, Jace Radke, Hector L. Ramos, Council Member Donovan Richards, Wilmer E. Sanabria, Erik Santana Medina, Gina Schrader, Greg Sehar, Mark Severts, Scott Sklar, Magaly Suárez, Ramòn Velasquez Pacheco, PJ Wilson, and Mike Winka.

Multiple other people contributed to this research in varied and crucial ways. Pace University Law School student, Robin Happel conducted a meticulous summer-long research project using her considerable talents and intelligence to set each chapter into context through humanities, science, and policy research. Karl R. Rabago, Katrina Fischer Kuh, and Dr. E. Melanie DuPuis all provided valuable advice and guidance on key points along the way. Many other Pace University departments also helped, including the Pace University Office of the Provost and the Office of Research, which aided the completion of this book with a research grant. I am grateful to my colleagues Dr. Avrom Caplan, Dr. Joan Walker, and Dr. Maria Iacullo-Bird, who were encouraging from the proposal stage and also helped access the practical means necessary to get the manuscript onto the page. Dr. Aditi Paul provided insightful input on early drafts of these chapters. The Pace University Energy and Climate Center, the Pace University Elisabeth Haub School of Law, and the Pace University Media, Communications, and Visual Arts Department were encouraging and supportive.

Without getting the interviews down, there would not have been a film or a book. I appreciate the Pace University film students who assisted during filming and post-production. Anna Wolfgram Evans worked remotely with me for an entire year during the completion stages, putting information in order and helping prepare these projects for public eyes. Anthony Parker, Gabriel Rivera, Irene Mercado, and Nicholas Ostrander all were skilled production assistants during filming and editing. I am grateful to the crew in each city, especially Leandro Fabrizi, Andres Otero, Toni de Aztlan, and Stephen Koss, who were incredible collaborators during filming.

My editorial team at Lexington Books, Nicolette Amstutz and Sierra Apaliski, were patient, accommodating, and incredibly helpful during the writing process. In addition, Johanna Campos, an extremely talented designer, created the cover that helped guide this book to you. Finally, several funders supported the film production which resulted in this book. They are: The Yip Harburg Foundation, The Puffin Foundation, Women Make Movies, and over 100 people who contributed to a crowd-funding campaign for the documentary film, "How to Power a City."

Acknowledgments

Thank you to each reader. By picking up this book, you are stepping past tired and outdated narratives about energy. By informing yourself through the lived experiences of people on the front lines of the clean energy revolution—and on the front lines of climate change and environmental justice catastrophes—you are taking the first steps to join them as a participant. They are each solving their community's energy problems, step by slowly gained and laborious step. They generously passed their knowledge on to me. Now, I am passing it on to you. I hope you will continue this forward cycle, pass this knowledge on to others, and continue transforming our energy system, one mind at a time.

Chapter 1

Introduction

This is not a traditional academic textbook. Instead, it is a virtual tour of the front lines of the clean energy revolution, and you will read firsthand accounts of how people around the United States and its territories brought—and are continuing to bring—clean, renewable energy to their communities. The following 11 chapters introduce the reader to nearly all aspects of electrical energy and how it governs our lives. They cover terminology from basic definitions to advanced concepts, technology and how these systems work, and the production and delivery of electrical energy. They also touch on city and state policies, economics and jobs, equity and environmental justice, how energy affects public health, land use and ownership of electrical production, and the design of clean energy systems. All chapters follow the one uniting theme: the transformation of our approach to making and delivering electricity.

WHAT IS THIS BOOK ABOUT?

This book looks at a variety of clean, renewable energy projects. The terms "clean energy" and "renewable energy" are somewhat interchangeable. In general, clean energy means producing energy from a source that does not create carbon emissions, toxic waste, or harmful particulate matter and does not contribute to air or water pollution. This term can also encompass nuclear power, as nuclear power plants do not create carbon emissions. "Renewable energy" means energy from infinite sources, like the sun and wind unlike coal, oil, or natural gas, which will eventually run out. For the purposes here, I use "clean energy" and "renewable energy" as umbrella terms for solar power, wind turbines, and geothermal, as those were the types of technology used in these case studies. This book does not look at nuclear power because of the design of the research. Neither is this book a complete and exhaustive overview of all types of clean and renewable energy. Instead, the following chapters describe how people in several cities brought, and continue to bring, renewable energy to their communities.

A clean energy project is getting easier to start and complete in some states and cities. However, that was not the case for nearly all of the people profiled in the following chapters. There projects were complicated. They were difficult and many required about a decade of attention, dedicated management, and a serious investment of time. Some projects were expensive, while others did not cost much and used donated equipment and labor. However, all saved money once installed.

Readers will also learn about something not discussed as often—the human aspect of this leadership. *Why* were the leader(s) of each solar, wind, or geothermal project motivated to take on these challenging projects? Some survived natural disasters; others wanted to prevail against man-made disasters. Still, others wanted to make their communities safer, save money, create jobs, and take part in addressing climate change. I pursued this research and have presented it in this form to advance conversations about the human aspect of energy transformation.

Readers of this book might be students in environmental sciences or environmental studies, sustainability, public policy, urban design, or law. Or, they might be doing interdisciplinary research or learning about an area where other fields intersect, like disaster management, climate change, or sustainable business. They might be energy professionals, or in a training program en route to becoming a solar, wind, or geothermal professional.

If you are reading this book, it is fair to assume that you have some interest in energy, clean air, and sustainability. Or that you are interested in making sure America's homes, schools, and businesses continue to have the reliable, consistent electricity that we all need. I hope this research provides insight into *how* these complex, challenging, labor-intensive projects can succeed and are succeeding. I also hope that the various clean energy projects described here will help you see how to guide your community to be part of transforming our electrical systems to be cleaner, more reliable, and safer.

CHAPTER SUMMARY

Each chapter addresses the methods I used to request the participation of the people involved. I used a narrative approach to this research and asked similar questions to most people. However, I also tailored interview questions to be appropriate. The narrative follows an emerging form of nonfiction called

Solutions Journalism, which focuses on describing the solution to a long-standing problem rather than detailing the problem itself. Each chapter also details additional research and the context for each project (see section below for more detail). Each of the subsequent chapters addresses a vital aspect of energy. Chapter 2, the Introduction, provides a brief and accessible history of energy and the grid, which is critical to understanding how we power cities. In chapters 3–10, I dive deeply into a specific city or town, a specific clean power project, and how the project addressed energy issues and other related issues. Finally, chapter 11 concludes all of these case studies and points toward directions for ongoing research.

METHODOLOGY AND DOCUMENTARY FILM

The bases of my research are interviews conducted for a documentary film called "How to Power a City." Each chapter looks at a specific way a community, family, or individual chose to use clean energy. The projects range in size from small emergency solar power systems that have a tremendous impact to sweeping legislation with the potential to transform energy use for a city with millions of residents.

Each chapter aims to elaborate on the diversity of energy fields, from gender, ethnicity, education levels, professional occupation, and public and private sectors. The people who shared their stories are from many walks of life: elected officials, activists, private citizens, business executives, lawyers, investors, entrepreneurs, and heads of utilities. Although the people portrayed in this book vary considerably in their approaches and have vastly different backgrounds, they all share one trait: they refuse to listen to the negative and incorrect narratives about clean energy. Or perhaps, they never believed them in the first place. Instead, they each took the initiative to find a way to bring clean energy to their home, city, or business.

The documentary, "How to Power a City," will be released in 2022 and will be an excellent companion to this book, as the film covers the same people and places in different ways. This book explores the narratives and include research, context, and academic resources, and the film adds visuals and explores a more personal side.

SELECTION OF PARTICIPANTS AND LOCATIONS

My methodology in selecting each particular location and type of clean energy project was to follow headline events and look for "evergreen" type stories, which maintain their relevance and currency for many years. Longevity is essential; the process of making a documentary and writing a book is lengthy, and each chapter needs to provide relevant information for years. In seeking people to go on camera, I sought guidance and advice from professionals, researchers, and others knowledgeable in the field and researched online for media stories about solar and wind projects. In addition, I sought out projects that would illustrate different aspects of electrical production. I reached out through email requests, typically followed by phone calls, then video interviews. For some cases, we did multiple interviews over a few years to track their progress; in others, there was only one interview.

In order to approach many possible people and projects, the following selection criteria were kept in mind:

1) Each situation used solar, wind, or some other type of clean power to provide energy for everyday use—it was not a lab, incubator, or test project;
2) The person showed a strong interest in their response to the initial email and a sense of enthusiasm for and commitment to communicating with the public about their work; and
3) Each project reflected a unique aspect of energy production, from city-level projects to off-grid emergency sized systems, projects led by utilities, and projects that challenged utilities.

Each chapter details a specific clean energy project which I researched and filmed for the documentary. I organized the book's text around these interviews, using key excerpts edited for brevity and clarity. Each chapter also sets the specific clean energy project into its location's time, place, and particular circumstances. Each chapter expands on the interviews with additional academic research, current news, and data related to the issues covered in the stories. In addition, many of the people portrayed in the documentary have been interviewed extensively in the media. Some chapters include quotes from these other interviews to expand the narrative. Readers who wish to learn more about a specific location or project can use these press interviews as resources for even more context.

INFORMATION GATHERING

In gathering this information, I used a narrative inquiry method, asking a series of descriptive questions about the history, beginnings, current plans, challenges, and origin of each particular project or initiative. I also followed the majority of the projects over at least a few months—only two of these were filmed in one interview—Vermont and La Riviera, Puerto Rico. After I completed filming, I continued to follow the project or work online, through email, and on social media for projects with social media accounts. For the most part, I avoided quantitative questions like "how much solar do you use? (by kilowatt or megawatt)" or "how much does it cost to build a wind farm?" These numbers are very site specific, require significant context, and are difficult to understand at an introductory level. More importantly, any numbers I used would be out of date when the first reader opens this book.

There were also practical reasons: numbers on solar and wind power vary wildly. They depend on how you measure the energy, who you ask, and what you are trying to understand. As mentioned, the lengthy duration of the creative process was also a consideration in today's highly dynamic clean energy markets. In fact, including a lot of quantitative data would have actually created a distorted picture, not an accurate one.

Instead, my approach was to ask about motivations, process, experience, hurdles, the day-to-day realities, and otherwise focus on the "*how*." I chose this qualitative look for creative reasons, as the interviews had to work well in a documentary film. By inquiring about *how*, each project came into being, I designed this book and the film in a way that I hope will stand the test of time. You will still find relevant numbers throughout this book, and there are dozens of easily accessible resources to find updated statistics in your ongoing research. However, there is a final and essential consideration: *we often depict energy as something to be summed up in a spreadsheet rather than by human experience.* As evidenced by the political battles over various types of energy production, energy transformation has cultural components with far-reaching and long-lasting implications. Clean energy transformation is about people as well as numbers. It is about our own understanding of how our energy systems function and the role we choose to take in this critical aspect of modern infrastructure.

The consequences of fossil fuel–based electrical production for our planet are well understood and widely documented, and there is widespread political and scientific agreement that we need to transition our electrical production to using clean and renewable technology. However, how we transition to clean energy is a picture that is still emerging. It is still up to individuals and communities to define what they want that picture to be. This book and the related film document this important side: the people and culture around clean energy transformation.

SOLUTIONS JOURNALISM

I follow a Solutions Journalism ("SJ") approach for both the book and the film. SJ is a rapidly expanding type of journalism adopted by hundreds of mainstream journalism newsrooms and embraced by writers, producers, and editors nationwide. Fostered by the Solutions Journalism Network, a nonpartisan non-profit founded by established journalists and writers who wanted to create a way to "help reporters, producers, and editors bring the same attention and rigor to stories about responses to problems as they do to the problems themselves." The tenets of SJ are to cover the way people respond to problems, explore the limitations of these responses, and examine related and ongoing issues. SJ is based on the philosophy that solutions stories can elevate public discourse, reduce polarization, and improve the overall quality of journalism and nonfiction, while maintaining journalistic accuracy.

In planning the structure of this book and the film, SJ provided a way to stay true to the narrative behind each clean energy project, represent the people and projects as authentically as possible, and focus on how each provides a solution. There is a large body of research on the benefits of clean energy production, how it works, what it costs, and what it can achieve. There is a much smaller body of research on *how* we can implement these projects, how clean energy projects address other critical local issues, and what clean energy cannot solve.

This means that I did *not* extensively cover the issues and problems related to fossil fuel energy production. In some chapters, I address serious problems such as air and water pollution through sections on related research. For most of the chapters, however, the focus is on creating the solar or wind project. There are volumes of studies on how using coal, oil, and natural gas to produce electricity creates greenhouse gases and carbon emissions, contributing to climate change. These studies also detail how fossil fuels and nuclear power pollute our air and water and create vast amounts of toxic and radioactive waste. By following an SJ approach, I organized this book to provide a new perspective and look at how communities move beyond fossil fuel use.

DIVERSITY AND EQUITY

The people portrayed in this book come from communities in six states and territories: Michigan, New York, New Jersey, Nevada, three locations in Puerto Rico, and Vermont. This diverse geography allows readers to explore clean energy from Caribbean islands to heavily forested mountain regions, deserts, dense cities, and coastal areas. Most of all, it looks at diverse people: women and men, from their 20s to their 70s, from various ethnic backgrounds, including African-American, Afro-Caribbean, Caucasian, and Latino. Education backgrounds also vary, with some people having extensive professional education in energy and others learning on the job. Some of the people portrayed in this book have advanced degrees from Ivy League institutions; others went to state or community colleges, have high school educations, or learned from community-based education. All achieved unique and impactful projects. Some people managed projects as part of their paid job; others started as volunteers in response to a community need. For some, those volunteer projects turned into professional jobs.

This diversity of gender, racial and ethnic identity, age, education, and professional affiliation is necessary. For too long, a very homogeneous group of people have dominated clean energy—typically with science or engineering backgrounds and typically male and white. Or, you might look for news on solar and wind power in the mainstream media, and end up reading the business section, where energy production is discussed in regards to investments, financing, and companies.

This approach needs to change. Energy affects everybody. Climate change affects everybody, although research has shown that the worst impacts of climate change, toxic waste, pollution, and fossil fuel power production affect communities of color far more than anyone else. Women lead clean energy efforts in dozens of roles, from scientists to business leaders to investors; yet, there is a yawning and acknowledged gender gap in the clean energy industry. Low-income families want just as much to be a part of the clean energy revolution as people with means; yet, they do not have the same access to tangible clean energy opportunities, such as financing for residential solar. There is a knowledge gap, and understanding things like state and federal tax incentives can be very challenging (if not impossible) for many people. Even if someone knows about financing or incentive programs, they might not understand how to access these programs nor have the resources to make an initial investment and buy the equipment.

To truly realize the promises of solar, wind, geothermal, and other clean energy, stakeholders who wish to see our energy systems transform to clean, renewable, affordable sources must also ensure that their work includes people from all walks of life. Building clean energy technology is not enough; we also need to develop clean energy public education programs, include diverse voices, showcase all types of leadership, and ensure everybody can benefit economically.

LIMITATIONS OF THE RESEARCH

One limitation of this book is that it does not offer full and complete solutions for every single one of the challenges listed above. This research project was grounded in qualitative analyses, and that means the results are not generalizable because of the limited sample size. However, these qualitative accounts demonstrate how people found creative and effective routes to progress, sometimes merely by refusing to give up, despite reasons to do so. My research discusses how each project moved beyond the bottlenecks and log-jams that often characterize the development of solar power or wind farms, and another limitation is that these solutions are not overarching prescriptions but rather examples of hyperlocal changes.

Another limitation is that each project still continues in real life, although the filming and interviewing concluded. I encourage readers to look up new developments in each location—Vermont, Nevada, Puerto Rico, New York City, Highland Park, and Atlantic City—and learn about the ongoing progress and challenges.

CONCLUSION

The discipline of filmmaking attempts to capture the person and story as authentically and clearly as possible so that the viewers can feel like they are experiencing the situation themselves, whether it is a conversation or an action. I tried to carry that sense of observation into writing this book and describe solutions, not prescriptions. The goal of SJ is to provide insight that helps others see the impact of new approaches while also highlighting the limitations of those approaches. This approach fosters the ongoing development of more solutions and supports progress in communities everywhere.

This is one reason new business models, new cultural approaches, and new ways to educate the public are as important as new technology. Innovations in the ways people can learn about, access, and use clean power will help address complex challenges of transforming our energy systems. We need to continue innovations like this, which open up opportunities for a wider variety of people to participate in the clean energy economy.

After traveling to 6 states and territories for this research, and interviewing over 30 people, I feel more motivated and am taking more action to bring clean energy to my own community. I switched to community solar, which I am lucky because it is available in New York City. I live in a large building with more than 150 apartments, and the community solar we buy covers the power we use in our apartment. Prior to learning firsthand about the many small starting points for the people and projects featured in this book, I might have believed things like a few apartments switching to community solar in New York City was too small to matter that much. The vast size of problems like fixing climate change and changing electrical production makes it easy to become overwhelmed about the possibility of transformation. Something like holding a "Solar 101" workshop or town hall about wind power for a couple dozen people in Queens, Salinas, Colchester, Highland Park, Atlantic City, or Las Vegas might seem like a tiny step. However, these stories show how tiny steps add up. The community solar program that my household uses went from 4 locations when I signed up in 2018, to over 10 sites about a year later, to 54 projects in 7 states 2 years after that. Starting with that small handful of New Yorkers, they have grown over 10 times in size.

At the time of writing, I am starting to work with others in my neighborhood to explore how we can use group purchases to lower the cost for rooftop solar and other sustainability technology and create a cluster of green buildings. I would have thought this is impossible had I not made this film or written this book. Small actions matter. Persistence does lead to progress. We have the technology, and now we need the human infrastructure. Environmental communications are pivotal to how people learn about, talk about, and communicate about renewable energy. This book and the related film are designed to contribute to ongoing research in environmental communications and related fields.

Each of these chapters shows the daily acts that constitute the front lines of the clean energy revolution. I now see every considerable achievement—an impactful piece of legislation, a solar installation, a wind farm, a community-owned renewable energy project with a battery backup—as a series of small actions. As you read through the stories, hopefully you will also see how significant impact began with small actions.

This sense of our own power is essential: *"It is an absolute necessity that we hold on to a sense of optimism in attempting to address climate change because otherwise it is all gloom and doom"* (Emphasis mine).[1]

It is also an absolute necessity that we understand that small steps like installing a half-dozen solar-powered streetlights, building emergency-sized solar power systems, and passing a city bill are not incomplete solutions. Instead, they are each a stride in a long relay toward a future where clean energy contributes to solving pressing issues from climate change to economic development, strengthening infrastructure, and improving public health.

NOTE

1. Bolstad, Jennifer. "How to Power a City." Interview by Melanie La Rosa. January 2018.

Chapter 2

The Front Lines of the Clean Energy Revolution

The Beginning

RESEARCH BACKGROUND

I am a lifelong environmentalist and have always been interested in solar and wind power but never really knew what to do about it. It seemed a very faraway thing, something wealthy people install on buildings featured in fancy architecture magazines. I had no way to understand how I, or anyone I knew, could be a part of it. Especially in New York City, where most people live in apartments and do not own our roofs. While New York City has a strong environmental community, living with sustainability in mind can take second priority to managing the exorbitant cost of living.

So when I saw this article, it was a jolt. Here is the quote that grabbed my attention: *"They cut their summertime electricity payments from more than $200 a month to the basic connection rate of $17, even as they run multiple air conditioners and power all the electrical gadgetry used by their two teenagers"* (Emphasis mine).[1]

From a Staten Island paper, solar advocates shared this article among sustainability networks in New York City. It described families purchasing their first residential solar power systems. For those who do not know what a "solar power system" might be, it is a series of solar panels that are mounted to a building, attached to an inverter, and goes into the main power grid. The electricity the solar panels make becomes a part of the mix of electricity flowing in that grid, while the rest of the power in the grid is from large power plants. When a family, like this one, buys or rents a solar power system, they get a reduction on their energy bill based on how much power the panels produce and how much electricity they use. The solar power systems described in this book are located on residential homes and businesses; they differ slightly from "solar farms," which are larger sites with dozens to hundreds of solar panels and not typically tied to one residence or business.

This family was 1 of 4 who were early adopters. Of course, what caught my attention was the part about their bill. Solar allowed these families to reduce their electric bills to the basic charge for the connection to the city's power grid. That fee was $17, or less than 10 percent of what the previous monthly electric bill had been before installing their grid-attached solar power system.

At the time, solar power systems were just starting to appear in private homes throughout New York City. As with most types of renewable energy, solar suffers from inaccurate and outdated stereotypes about its efficacy, reliability, and ability to produce enough power for a modern lifestyle. This article was shared widely as evidence of what solar advocates had known for a very long time: even in New York, which is far from the sunniest state in the country, you can produce enough solar power to run an average home.

Around the same time, another type of clean energy—wind power—was showing up as small wind turbines on New York rooftops. Many of these were luxury condo buildings, and these small rooftop wind turbines are not the same as the typical type, a tall pole with three large blades. These wind turbines were barrel-shaped, with a vertical axis that allows them to work on a rooftop.[2] The visibility of the turbines drew attention: a grocery store in Brooklyn installed several in their parking lot. A large condo in Long Island City has three. They added to the unique contours of the New York City skyline.

I started researching how all of these pieces work together, and tried to understand how I could be a part of this clean energy movement. I realized I could not learn about clean energy technology without learning about the grid, how electric system works, and how clean energy technology is part of a much bigger system. I just wanted to be able to use solar or wind power myself, and see if, maybe, I could be part of a movement helping our country to convert to clean energy as a way to address climate change. At the time, I had no idea about the need to survive hurricanes using resilient systems, the improvements in design offered by solar and wind systems, what distributed generation meant and why it is important, or the economics of the energy marketplace.

Through the process of making a film, I learned about all of these. The interviews I did with the people portrayed in this book and the film went far beyond solar panels and wind turbines. They were a detailed education about power plants, solar and wind power marketplaces, the grid, the difference between on-grid and off-grid systems, regulation by state governing bodies, and many other essential concepts related to clean energy.

CHAPTER SUMMARY

This chapter introduces a variety of concepts and relevant vocabulary and is a necessary baseline for subsequent chapters. First I discuss briefly, how our electricity system works, the way it was explained to me. Then I discuss how this system works currently and how we cannot talk about solar and wind integration without talking about the grid. Finally, I have included a glossary-type section discussing much of the terminology, as many concepts discussed later in this book require the reader to know these terms. Hopefully, this helps make these complex systems understandable for people just learning about clean energy, while still being of interest to those with more experience.

A BRIEF AND ACCESSIBLE OVERVIEW OF OUR ELECTRICAL SYSTEM

The Bathtub Metaphor

Early in my research for the film, I ran into some people who had set up a card table outside of a food co-op in Brooklyn and were asking people to switch to wind power on their electric bill. They had beautiful photographs of upstate New York wind farms. I asked them a basic question: "if the wind farm is upstate, how does the electricity power my apartment?" My understanding of renewable energy system was determined by rooftop solar; you put the solar panels on your roof and the panels power your electrical devices. Except that is not how urban rooftop solar actually works, such as the systems on the Staten Island homes.

The guys at the card table gave me a kind of funny look and did not answer the question. They kind of mumbled something about the power company and shuffled their flyers, asking me again if I wanted to sign up. I did not know if this was because they did not know the answer, or because that question is, for people in the energy field, very basic. The looks on their faces suggested the former. But I realized I had a lot more to learn about how renewable energy worked.

For those who took multiple science classes, or have advanced knowledge about electrical systems, might want to skip this section. But for those who are at a fairly introductory level or really don't understand how our electrical grid or a renewable energy system works, this section covers the basics.

I started asking around and someone explained a simple and revealing concept: the bathtub metaphor.[3] All electricity produced, whether from a coal-burning power plant or a solar panel on the roof of an organic bakery, goes into one giant combined system. Just like having multiple taps filling one big bathtub. Some of the water might be dirty, making the water in the tub dirty.

But as you turn on new taps with clean water, the water in the tub eventually becomes clean. The faster you turn off the dirty water taps and turn on the clean ones, the faster it gets cleaner. Once that electricity is in this "bathtub" in the form of electrons, it is all the same. There is no difference between the electrons created by solar, wind, oil, natural gas, or other sources. But there is a tremendous difference in the amount of pollution created during the making of these electrons.

The bathtub is our grid: the electric lines that stretch from house to house, up and down our blocks, along highways. The grid marches through mountains and deserts and fields in giant high-voltage transmission towers. You've seen this grid so often that you likely do not even see it at all, as it has blended in as a normal part of the landscape. In recent years, we are starting to see regular grid collapse from storms and periods of extremely high use which overload the grid. Because of these grid collapses, we have started paying much more attention to the grid.

Nearly all homes in the entire country are connected to the grid. Unless it is an "off-grid" house, which are typically located in rural or wilderness areas where the grid does not reach, buildings are connected to the grid. Buildings must follow local codes, city and county ordinances, and other laws, which do not make it easy to build an off-grid home. One-hundred percent electrification of our country was a major advance of the 19th and 20th centuries, and being off-grid has traditionally been the realm of survivalists.

As I continued filming, I learned another fundamental concept from a very smart engineer who realized the gap in my understanding. It is typical electrical systems deliver electrons to your home the instant they are produced. Storing electricity to be used later has been one of the biggest challenges to using solar and wind power.[4] While the bathtub metaphor is a great way to start understanding the grid, unlike a giant tub, which is a reservoir that can hold water indefinitely, electricity needs to be used the instant it is created. Perhaps I just did not pay enough attention in my middle-school science classes, or perhaps they never taught this, but this one quite basic fact completely changed my understanding of the grid-based system and its challenges to distributing renewable energy. No one needs a middle-school science class to understand that solar panels produce the most electricity in the middle of the day and that wind turbines create the most electricity when the wind is blowing hard. However, we need to understand that when those types of power production put electrons in the grid, those electrons need to be used immediately. Just as putting too much water in a bathtub will make it overflow, putting too much electricity in the grid can cause explosions. Electricity produced by solar power systems or wind turbines has, up until very recently, needed to be used immediately to ensure it did not cause dangerous conditions on the power grid. If that house in Staten Island produced a lot of electricity on a blazing day in August, and did not use it all immediately, because it was attached to the big bathtub (the grid), those electrons would go to the home of someone else on the grid. Too many solar panels putting electrons in the grid on that same August day could, at one point, have caused a power surge, although this is an aspect of solar power integration that has largely been solved by most utilities.

Now that battery technology has come into use for utilities and homes, we have bigger bathtubs that hold the energy until it is needed. In subsequent chapters, we will come back to this because the chance of "overflow" is being addressed through battery technology—often referred to as "energy storage"—that can go on residential homes, businesses, and utility sites.

Grids and Microgrids

That same engineer suggested reading "The Grid" (Bakke, 2016) as an excellent resource for a deeper understanding of how the power system works.[5] It describes another critical concept: that the first power companies developed the grid to take electricity to scale, and make it affordable enough for a regular family to buy it. In the very early days of electricity, there was a moment when wealthy families might have built their own small-scale power generators for their farms, businesses, factories, and large homes. Families who could not afford to install generators would have remained without power. The development of a grid by investors like Edison and Westinghouse allowed the average family to buy electricity. This was done for purposes of capitalism rather than democracy; however the effect was that all homes and businesses could access electrical power. Buildings became tied together through an electrical grid and achieving 100 percent electrification became a mark of a developed nation.

This grid has grown to massive sizes, and the mainland U.S. now has only three power grids: one in the East, one in the West, and a separate one in Texas. Islands like Puerto Rico and Hawaii have unique systems, and because they have discrete grids, there are also unique ways in which islands can transform their energy stems.

This historical look is important because now, some places are looking for ways to separate from these three main grids. There are many reasons for this, including being able to use 100 percent clean energy for a community, home, or business. Another reason is to prevent widespread blackouts—while being interconnected means that our electricity can be cheaper, it also means large sections of the grid must shut off if one part of the system is down. An energy investor in Puerto Rico explained it succinctly: "if it is broken anywhere, it is broken everywhere." Clean energy allows for something called a "power microgrid" in which a building or other entity, for example, a hospital, separates from the main grid. Energy professionals also refer to this as "islanding off" or a "power island." In the event of a large hurricane, being able to island off hospitals and other key infrastructure and maintaining power is a critical need. It is important to note that any type of power production can create a microgrid or power island. It does not have to be solar or wind power. In fact, microgrids are often created on hospitals, military bases, universities, and large residential complexes using gas generators. A solar power array and battery are an ideal way to create a microgrid, and are one of the energy projects I describe in this book.

CLEAN ENERGY ON THE GRID

Once you understand this idea of a big reservoir of electricity that needs to be used immediately, then you can understand how a home in Staten Island goes from a $200 electric bill to a $17 one—and can still use their TV, blow dryers, gaming stations, air conditioner, and all other devices at any hour of the day whether the sun is up or not.

The solar panels on that house, or any other home or business, create electrons. They feed into that giant grid, along with all of the electrons from nuclear power plants and oil, natural gas, and coal-burning power plants in Staten Island, Brooklyn, Queens, Manhattan, Westchester, and Long Island. Those solar panels create only a tiny drop in that big bathtub—just like that families' power use is a very tiny draw from the bathtub. But it adds to the system. That family just goes about their normal lives, using electricity whenever and however they like. The New York utility that manages that grid, tracks how much power the Staten Island family's solar panels contribute to the grid. They also track how much the family uses. If their solar panels make more electricity over the course of the year or the month, than the family uses, the family does not pay for electricity. This is referred to as "net zero." While there are times when the family draws from the grid, such as nighttime and cloudy days, as long as they put electrons into it, they maintain net zero use. For this to work, a billing mechanism called "net metering" keeps track of the family's use compared to how much power their solar panels produce. Utilities in states with net metering programs reimburse families for the electricity they produce. These rates that utilities pay for net metering vary greatly, and can become very controversial. Despite this, every report I read and professional I spoke with regarded net metering as one of the most effective ways to integrate solar power while also providing a way for families to save money.

Net metering works differently in different states, based on state and utility policy, and is one of the core reasons the residential solar market has grown tremendously in the past decade. Net metering programs allow a household to generate a credit with the utility when they produce a lot of power in the summer. Then, when they draw a lot of power from the grid, in winter or at night, their bills are reduced based on the credit on their account.

The same general concept works for onshore wind farms which are privately owned, except on a bigger, more business like scale. The wind farm makes electricity, which is sold into the grid, typically through something called a "power purchase agreement." If the wind farm is on a private land, for example, a ranch, that landowner is often paid for the use of that land. In some cases, a municipality receives payment for clean energy projects; for example, Lowell, Vermont, the power company compensates the town for a nearby wind farm. In my research, many people discussed how power purchase agreements are critical for large-scale development like wind farms, which create vastly more electricity than residential solar systems.

At the time of this article about the Staten Island family, in the 2010s, managing this input to the grid from various sources was a challenge for many utilities. Traditionally, the grid was built to flow in one direction: from a power plant to a home or office. Just like a household cannot put water into the faucet, at that time, putting electricity into the grid from one of these sources created a technological hurdle. Another technological hurdle was, as mentioned above, power flowing from solar or wind installations could cause a swell of electrons in the middle of the day or when the winds were strong. Back to the metaphor of the overflowing tub, utilities either could not or did not know how to manage these swells of electricity. Surges can be dangerous, can overwhelm the capacity of the lines, cause explosions, and take down the entire grid. At the time of writing this, a large number of utilities have resolved these issues, found ways to accept electricity into the grid, and developed systems to manage renewable energy. There are many resources for utilities that have not developed this technology yet, and that want to catch up to their peers in other states. Utilities also developed state organizations that control this back-and-forth flow, called "Regional Transmission Organization (RTO)" or the "Independent System Operator (ISO)," which function like a clearinghouse or central management point, monitoring all the electricity flowing into the grid to prevent these explosions. Since electrical use on a large scale is actually quite predictable, it is possible to know when solar power or wind power will increase or decrease and add or subtract other sources as needed.

States which are successfully integrating clean energy in large amounts have utilities that embrace this type of "mixed generation," or power production from many sources. They also support "distributed generation," or making power on rooftop solar systems, wind farms, large solar fields, and a variety of sources distributed throughout the utility region, rather than being in one centralized power plant.

Opposition to Clean Energy

One of the most common points of opposition to renewable energy is that the "sun doesn't always shine and the wind doesn't always blow." This common adage expresses what some view as the limitations of wind and solar power. Like most adages, it is a little bit true but conveys an entirely wrong meaning. While in a technical sense, it is accurate. Again, no one needs a middle-school science class to know that the sun sets every day and winds calm down eventually. But the underlying meaning of this saying is to undermine the trustworthiness and reliability of renewable energy systems. This is inaccurate. The renewable power system works. What does not work is the design of the system, and we have not yet fully developed and marketed systems that work well with intermittent production.

Our electrical system was created over one-hundred years ago, and designed for what was possible at that time: to make electricity in a power plant by burning fossil fuel, which creates steam, turns a turbine, and creates electricity to be sent through wires to a building. Now, we have the technical capacity to produce electricity in various sites and use different means of production. Rather than dismiss solar, wind, and other types of renewable power because they do not work in the same way as fossil fuels, we need to develop technology that can work with these intermittent systems. The clean energy technologies explored in this book are examples of reconfigurations of our power system. This more appropriate response harnesses rather than dismisses the proven ability of clean energy sources. In fact, in recent years, we have seen how unreliable and vulnerable our current grid-based systems are. Massive blackouts are becoming more and more common, caused by everything from unprecedented hurricanes to something as simple as a tree falling on a wire.

To update that adage to be more accurate, the sun shines every single day, in very predictable ways. In fact, we have reams of data on the hours and consistency of the sun in every state and territory. The wind also blows in similarly well-documented ways; in many areas like coastal regions and out in the ocean, wind is extremely predictable. So, the adage should really be: "the sun shines every day, there is always wind blowing somewhere, and the sooner we transform our energy systems, the more we can use clean, consistent, affordable, and reliable power."

TERMINOLOGY

A great deal of terminology is used throughout this book for energy production, distribution, types of energy, and business models. This is a short list of some of the relevant terms, which will be helpful in understanding the rest of this book. Please see the index at the end of this book for an even fuller reference.

Clean Energy—Clean energy sources do not produce greenhouse gases, and typically mean solar, wind, and hydropower, but also might refer to nuclear power. Often used interchangeably with "renewable energy."[6]

Community Solar—Definitions vary. According to the U.S. Department of Energy, National Renewable Energy Laboratory (NREL), community solar is "a solar-electric system that provides power and/or financial benefit to multiple community members."[7]

Demand—How much electricity a system, building, or piece of equipment uses. Electricity demand varies based on factors like population, amount of industry, weather, and patterns of human activity. For example, on summer days when everybody turns on their air conditioner, demand is sky-high. On the other hand, demand is low in the middle of the night when people are sleeping and businesses are closed.[8]

Distributed Generation—Energy generation at or near the source where it will be used, typically solar panels, combined heat and power, wind farms, and other types of renewable energy. Distributed generation can power a home or business or be part of a larger microgrid.[9]

Duck curve—The difference in electricity demand and the amount of available solar energy throughout the day, with a large rise in solar power in the middle of the day, which wanes in the evening when the sun sets. It somewhat resembles the curve of a duck's back.[10]

Energy Storage—A way to store electricity on the scale needed by a home or utility. Used synonymously with batteries.[11]

Grid attached—A power system attached to the main grid.[12]

Islanding and *Island Off*—When a distributed generation source is disconnected from the utility grid, and switches to a microgrid.[13]

Kilowatts, Megawatts, and Gigawatts—The way the size and productivity of a power plant of any size are measured. Kilowatts (kW) are smaller in scale; the average residential solar PV system is approximately 5kW, while a 10kW system is considered the size for a large home or small office. A much bigger system like 1 megawatt (MW) is on the scale of a small solar farm or for industrial use, or a small wind farm might be around 30 MW. A gigawatt is a utility-scale size; the largest solar farm globally, currently in India, is 2.2 GW.[14]

Load—How much electric power is delivered on a system, or required by that system, based on consumer use.[15]

Microgrid—A smaller grid connected to the main electrical grid.[16]

Net metering—A billing mechanism that credits solar energy system owners for the electricity they add to the grid. Specific policies and rates vary from state to state.

Off-grid—A power system that is not attached to the main grid, and functions as a stand-alone system.[17]

Peak times or peak demand—The maximum load during a specified period of time.[18]

Power purchase agreements—Agreement to purchase the power from a renewable energy or other source. Some residential solar systems use this term to refer to a leasing arrangement for rooftop solar.[19]

Renewable Energy—Renewable resources include solar and wind power, or other types of power production that do not deplete when they are used. Can include sources such as biomass and geothermal which might have emissions. Often used interchangeably with "clean energy."[20]

Solar Panel Output—How much electricity the solar panel system makes, which depends on siting, days of sunlight, and the efficiency of the panels.[21]

Solar PV or PV—Solar photovoltaic, a solar-powered system for creating electricity.[22]

Solar thermal—A solar power system for producing heat, as opposed to electricity.[23]

CONCLUSION

Energy production is both science and engineering and policy and politics. Some of the following case studies focus more on science and engineering and showcasing the "how" of clean energy. Others focus more on the policies and politics that intersect with science and govern the engineering of electricity. One of the newer insights this book explores is how clean energy infrastructure, when engineered properly, can protect our communities from the worst impacts of climate change, such as devastation from hurricanes, as well as reduce our contributions to climate change.

How can we similarly continue to apply science and engineering to city, state, and federal policies to create energy infrastructure that protects us? How do we apply clean energy to build infrastructure for a city that reduces its climate impact? Each of these chapters illustrates some excellent ways to begin.

NOTES

1. Young, Deborah. "Sun's the Secret to $17 Electric Bill." *Staten Island Advance*, originally posted March 14, 2010, updated January 03, 2019. https://www.silive.com/news/2010/03/suns_the_secret_to_17_electric.html

2. Chaban, Matt A.V. "Turbines Popping Up on New York Roofs, Along with Questions of Efficiency." *New York Times*, May 26, 2014. https://www.nytimes.com/2014/05/27/nyregion/turbines-pop-up-on-new-york-roofs-along-with-questions-of-efficiency.html

3. Brooks, David. "More on That Metaphor About the Electric Grid as a Bathtub." *Concord Monitor*, November 14, 2017. https://www.concordmonitor.com/electricity-power-grid-13516204

4. Bolger, Kyle. "How To Power A City." Interview by Melanie La Rosa. March, 2018.

5. Bakke, Gretchen. *The Grid: The Fraying Wires Between Americans and Our Energy Future*, 1st edition. Bloomsbury USA, July 26, 2016.

6. UCLA Luskin Center for Innovation. "Progress toward 100% Clean Energy in Cities and States Across the US." *Luskin Center for Innovation*, November 2019. https://innovation.luskin.ucla.edu/wp-content/uploads/2019/11/100-Clean-Energy-Progress-Report-UCLA-2.pdf

7. National Renewable Energy Lab. "A Guide to Community Shared Solar: Utility, Private and Non-profit Project Development." *National Renewable Energy Lab*. U.S. Department of Energy. May 2012.

8. Blumsack, Seth. "Introduction to the Electricity Industry: Electricity Demand and Supply in the United States, EME 801: Energy Markets, Policy, and Regulation." Penn State, *Department of Energy and Mineral Engineering*, website accessed December 12, 2020. https://www.e-education.psu.edu/eme801/node/490

9. U.S. Environmental Protection Agency. "Distributed Generation of Electricity and its Environmental Impacts." *Epa.gov*, accessed December 29, 2020. https://www.epa.gov/energy/distributed-generation-electricity-and-its-environmental-impacts

10. Jones-Albertus, Becca. "Confronting the Duck Curve: How to Address Over-Generation of Solar Energy". *Energy.gov*, October 12, 2017. https://www.energy.gov/eere/articles/confronting-duck-curve-how-address-over-generation-solar-energy.

11. SEIA. "Residential Consumer Guide to Solar Power."

12. U.S. Department of Energy. "Grid-Connected Renewable Energy Systems." *Energy.gov*, accessed December 29, 2020. https://www.energy.gov/energysaver/grid-connected-renewable-energy-systems

13. Jain, Monika, Sushma Gupta, Deepika Masand, Gayatri Agnihotri, and Shailendra Jain. "Real-Time Implementation of Islanded Microgrid for Remote Areas." *Journal of Control Science and Engineering*, April 18, 2016. https://www.hindawi.com/journals/jcse/2016/5710950/

14. SEIA. "Residential Consumer Guide to Solar Power." Solar Energy Industries Association, 2018. https://www.seia.org/sites/default/files/2018-06/SEIA-Consumer-Guide-Solar-Power-v4-2018-June.pdf

15. Retail Energy Supply Association. "Energy Glossary." *Resausa.org*, accessed December 29, 2020. https://www.resausa.org/shop-energy/energy-glossary#r.

16. U.S. Environmental Protection Agency. "Distributed Generation of Electricity and its Environmental Impacts."

17. U.S. Department of Energy. "Grid-Connected Renewable Energy Systems."

18. Retail Energy Supply Association. "Energy Glossary."
19. SEIA. "Residential Consumer Guide to Solar Power."
20. UCLA Luskin Center for Innovation. "Progress toward 100% Clean Energy in Cities and States Across the US."
21. SEIA. "Residential Consumer Guide to Solar Power."
22. SEIA. "Residential Consumer Guide to Solar Power."
23. U.S. Energy Information Administration. "Solar Thermal Power Plants." *EIA.gov,* website accessed November 2, 2020. https://www.eia.gov/energyexplained/solar/solar-thermal-power-plants.php

Chapter 3

Atlantic City, NJ
Offshore Wind and Mixed Generation

INTRODUCTION

On the wind-swept South Jersey coast, is the Atlantic County Utility Authority (ACUA), a 24/7 waste treatment plant running entirely on clean energy. With five wind turbines, ACUA's profile is unmistakable in the marshlands outside Atlantic City, where the reliable coastal winds keep the water pumps working all day and night. Other parts of ACUA's energy systems include a massive solar carport, methane recapture system for landfills, and off-grid batteries. Combined, these make ACUA a mixed-generation clean energy mecca.

Inspired by their success, a group of commercial fishermen based in the Atlantic City region proposed building what would have been America's first wind farm. The idea attracted widespread local support and won a $50 million Department of Energy grant. Despite this, Fishermen's Energy applied three times for approval of their plans by the New Jersey Board of Public Utilities. They never received it. Had they been successful, they would have been the first offshore wind farm in the United States.

Offshore wind is planned and expected to grow considerably over the next decade. All of the Atlantic states from Maine and Massachusetts, through New York, New Jersey, Maryland, Virginia, and the Carolinas, have set high goals for offshore wind power generation. However, at time of writing this, only 2 U.S. states have an offshore wind farm—Rhode Island was the first, opening the 5turbine Block Island Wind Farm in 2016. 4 years later, Virginia followed with two fully functional turbines that became operational in October 2020 as a pilot for the much larger 180-turbine dominion wind farm. The public plans for the entire wind farm slate construction to begin in 2024, although ongoing contracting, permitting, and other types of review are still required.[1] The Pacific states are also planning offshore wind development, which will unquestionably be a major area of energy development in the coming decades.

CHAPTER SUMMARY

This chapter explores the potential of wind farms, the technology and business aspects of starting a wind farm, and the perils of navigating the permitting process, which is very different than a solar farm.

This chapter focuses on the technology of how the ACUA's various clean energy technologies work in concert with each other, as well as how Fishermen's Energy was planned, their process to seek state approval, and the challenges they met along the way. This chapter is not an exhaustive look at offshore wind nor a prescription for managing the complex development and approval process. Instead, this chapter follows two particular examples for the insights they provide. The first is the ACUA, an industrial site regionally renowned for its early adoption of clean energy, mixed generation from clean energy sources, and commitment to educating the public and others in the industry about clean energy. The second is Fishermen's Energy, which almost became the first offshore wind farm in the U.S., intending to bring jobs to South Jersey and develop the American offshore wind industry.

I believe this chapter provides readers with insight into clean energy development, particularly in looking at how our cities and states will rise to meet renewable energy goals. Any discussion of clean energy would not be complete if it did not discuss wind power and look at offshore wind and its massive capacity for power production.

METHODOLOGY

The ACUA was the first place I reached out to in researching for the documentary. Since my research started with people at a sidewalk table talking about wind power, I wanted to find the closest wind farm to New York City. That turned out to be the ACUA. When I learned that they also had a parking-lot-sized solar carport, other types of clean energy, as well as wind turbines, they seemed an ideal location to start exploring what it would mean for our cities to turn to clean energy. Monica Coffey at the ACUA was incredibly helpful in setting up interviews with Rick Dovey and arranging for me to be at their site during one of their regular public tours. When I interviewed Rick, he put me in touch with Jeanne Fox, a widely known and respected leader of clean energy programs for the state of New Jersey. Jeanne Fox suggested I speak with Paul Gallagher at Fishermen's Energy to learn more about the offshore wind farm they were trying to build. I first interviewed Paul in 2015 and followed Fishermen's Energy through the end of 2018 as they sought approval from the New Jersey Board of Public Utilities. Along the way, I attended two large wind power events, Time for Turbines, held at the ACUA, and the International Partnering Forum for Wind, an annual conference hosted by the Business Network for Offshore Wind. These events provided insight into the broader world of offshore wind development and the many stakeholders necessary for the U.S. to develop an offshore wind industry.

OFFSHORE WIND IN THE UNITED STATES

When I met Paul Gallagher at Fishermen's Energy, they were poised to become the first offshore wind farm in the United States. New Jersey would have had the distinction of being home to the beginnings of the offshore wind industry in the United States. I began filming with Fishermen's Energy in 2015. Block Island Wind Farm came online in December 2016,[2] beating Fishermen's Energy to the number 1 spot.

To understand the context, I followed wind energy closely through news alerts. I was extremely surprised to see how many news stories would publish about the "next" offshore wind farm, given the fairly slow pace of offshore wind in the U.S. While every state on the Atlantic coast is conducting research and development on offshore wind, and leasing rights for wind exploration in the waters off of their shores, there are still only two functional offshore wind farms.

I followed these stories, sure that soon enough, I would point my camera at an offshore wind farm. That I would film a crew of technicians building the bases, the towers, and installing the turbines, and would interview a manager on how the turbines function, what kind of maintenance they need, how they affect the ocean, and their generating capacity. Maybe I would even interview scientists studying the impact on ocean life and how to build the turbines to benefit ocean life and coastal communities. At the beginning of any documentary, when you commit to following it to where the story leads, you still anticipate arriving at certain turning points. The anticipation for this film was that I would, soon enough, start speaking with members of the public at events to open the wind farms. And at these events, I would conduct interviews sharing the excitement of those eager for this new industry and learn about the fears of those opposed to it. For audiences of the documentary, it would be a chance to witness the birth of an industry ripe with opportunity. But none of that happened, as the duration for offshore wind development is even longer than the lengthy process of making a documentary.

Filming did not ultimately include the unveiling of offshore wind turbines—a visual that would signal the milestone moment of offshore wind's arrival in the United States. However, there is a healthy market for onshore wind farms development, and the two existing offshore projects mentioned above. By the time this book is in your hands, there will possibly be far more offshore wind farms. I certainly hope so.

I believe that what I ended up filming is the actuality of the birth of a new industry: financing, planning, partnership building, and seeking permits and approval. That industry is born of investment of time, money, and personal energy—the moment of cutting a ribbon might be good footage, but it is most definitely not the actual birth of an industry. I had hoped to see Fishermen's Energy follow the example of the nearby ACUA, both smaller companies establishing innovations in their field. Instead, I ended up with two very different portraits of business development and a glimpse at how immense, how much potential, and how important the offshore wind industry will be in the coming years.

Wind energy researchers in 2015 commented that "the United States is perhaps further from commercial-scale offshore wind deployment today than it was in 2005. Meanwhile, Europe went from 622 megawatts of offshore wind capacity in 2004 to more than 8,000 in 2014 across 74 wind projects, with those under construction to increase capacity to almost 11,000 megawatts."[3] Particularly with the lofty national and state goals for clean energy, including offshore wind, ongoing research on all aspects of offshore wind development has the potential to ensure it becomes a key technology in America's energy mix and not a missed opportunity.

My interviews revealed many challenges. Some were much unexpected: for example, when I was filming with Fishermen's Energy, the blades for offshore wind turbines were only manufactured overseas, adding significantly to the overall project cost because of the shipping charges. In addition, there were abundant environmental concerns, ranging from the potential impact of offshore wind turbines on ocean life to how they would affect the migration of large marine mammals and whether the turbine bases were actually beneficial by providing habitat for fish. All of these merit serious, well-funded, ongoing studies to ensure that large offshore wind farms have the least negative impact possible. Other issues that came up repeatedly were the need to train local maintenance crews, make sure that offshore wind development benefited local economies, and develop onshore infrastructure that ties the offshore turbines to the onshore power grid—all of these merit ongoing study and collaborations to ensure that the economic benefits extend to everyone.

Potential to Lessen the Impact of Hurricanes

Another theme throughout this book is the potential for clean energy technologies to help communities survive the worst impacts of climate change and reduce emissions that are causing climate change. Throughout this book, I look at the capacity of solar power to come online immediately after hurricanes, particularly for rooftop-mounted systems with batteries or when there is a microgrid. Researchers in wind energy are also finding that offshore wind turbines can help mitigate hurricane damage by slowing down winds and currents if the turbines are sited correctly.[4] Using computer modeling, these researchers have shown that large turbine arrays have significant potential to lower storm surges.

This chapter is set in a region of New Jersey that was battered by Hurricane Sandy in 2012 and was without power for lengthy periods. While researchers currently use computer models, the potential for offshore wind farms to protect communities from the worst of storm surges, such as those during Hurricane Sandy, is another area that merits more ongoing study. As an important consideration for resiliency, this also seems like an essential aspect of offshore wind for public conversations. Hurricanes Irene and Sandy both devastated New York and New Jersey from the coasts to the mountains. With this level of devastation quite possible from future hurricanes, how would the public opinion of offshore wind be different if they knew that it might prevent or lessen hurricane damage? As with other clean energy technologies discussed in this book, this lesser-known but highly impactful aspect deserves attention in public conversations about clean energy.

ATLANTIC COUNTY UTILITY AUTHORITY

Rick Dovey is the President of ACUA. He is dedicated to clean energy and smart business and has led ACUA to regional and international recognition. ACUA might be the only wastewater treatment plant in the entire country that hosts public tours. It is, after all, a place to treat sewage and can have a very strong smell. Despite this, ACUA hosts regular public tours, as people want to see the turbines up close. I filmed on a day where a group of approximately 25 people were touring the site. Families with children, international visitors, and people taking day trips—all to see the turbines. ACUA is visible from the Atlantic City casinos, and another anecdote that Rick shared with me is that the casinos would have people come in asking for a room with a view of the wind turbines. This goes against the prevailing narrative that the public hates wind turbines and believes they destroy ocean views.

Rick explained that ACUA operates 24/7, and because of that, their power costs were hitting the roof during the energy crisis of the early 2000s. Facing skyrocketing bills, they accepted a proposal from a wind energy company to be a test site for a small onshore wind farm—the Jersey Atlantic Wind Farm—which was built in 2005. Later they added a solar-covered carport, methane recapture from landfills, geothermal, compressed natural gas-powered vehicles, and electric car charging ports (see Figure 3.1). Their website has extensive information on all of these, including a portal to monitor the power output from the turbines and solar panels.[5]

He provided a bit of background on their motivations and results:

> *"The projects that the ACUA has put into place, which have been incorporated into our operations, have changed the focus of our organization to not just any wastewater and solid waste authority. You're always looking for an identity, this gives us a very progressive, forward-thinking, environmentally progressive feel with our community. The people we serve know who we are. We're part of them"* (Emphasis mine).

Figure 3.1 Mixed Generation of Wind and Solar Power at Atlantic County Utility Authority. *Source*: La Rosa, Melanie, 2017.

Over a decade after being installed, these five turbines produce 60 percent of the power that ACUA uses. The rest comes from solar, a landfill methane recapture project, and net metering. In that time the wind farm has saved ACUA almost $5 million in power costs. Rick and his colleagues skillfully navigated a vast minefield of policies, regulations, and public and private funding.

ACUA is a highly functional showcase for multiple types of clean energy technology. As batteries and microgrid technology became more available, particularly in the few years after Hurricane Sandy, Rick and the ACUA staff added a utility-scale battery project. They continue to integrate emerging clean energy technologies to their site and have established themselves as a regional leader in Environmental Protection and clean energy.

Two points of opposition often come up with wind: people feel like they destroy views and the noise the turbines make. For the second, the ACUA's industrial site is already so noisy from the pumps and the wastewater pools, the screeching seagulls, and the constant wind blowing that I barely noticed any noise from the turbines. If I stood underneath one, I could hear a subtle "swoosh." For the first, there is Rick's story about the casinos, the visitors who come with their kids, and this story, shared by Rick:

> *I'll give a story that just about a week after the wind farm came online, I was at the Lowe's on the weekend. I saw down the aisle a couple that I had known from coaching soccer. The woman came running down towards me, saying, "the wind farm—it's so great! Thank you!" She came and gave me a big hug. So . . . not everybody gave me a hug, but people were proud that this was happening and had happened in their city and their location. And that it was a local agency that was a part of bringing it to fruition.*
>
> *I see that in Venice Park, the neighborhood across the highway, and the closest group of homes to this wind farm. The folks there think of this as their wind farm and are very proud of it. They'll say, "Oh, I live near the wind farm." Ok? So, it's a positive thing. It's not like, "Oh, I live near the power plant."*

It is worth detailing that the Venice Park neighborhood Rick mentions, which is the closest to the wind turbines, fits all the criteria of an environmental justice community. It was one of the first Atlantic City neighborhoods where African-Americans could purchase homes outside of redlined areas. Today, it has a diverse population, with close to 40 percent of families living below the poverty line.[6]

Venice Park has also experienced a rise in periodic flooding due to coastal surges and tides. Located along the Bay, along with surrounding communities, it was built on filled marshland. During 2012's Superstorm Sandy, massive flooding all along the Jersey coast exposed the vulnerability of all coastal communities in New Jersey, with many of the most at-risk communities populated by African-Americans, Latinx, and immigrant families.[7] A 2019 National Oceanic and Atmospheric Administration study predicted that annual flood days in Atlantic City will increase from nine days in 2019 to between 20–35 days in 2030, and 65–155 days by 2050—an increase that could mean ten times as much flooding for families with the least amount of resources.[8]

In that context, the wind farm located next to Venice Park is the converse of the usual way energy systems are situated. A low-income, diverse neighborhood lives next door to an energy source that is clean, renewable, and reduces pollution. An energy source that will, if taken to scale, help mitigate the climate change that causes the flooding. Not only does this speak to the immediate needs of the community to have reliable and affordable electricity, but it also speaks to something intangible yet powerful. Seeing cutting-edge technology visible outside one's doorstep is, as Rick notes, a way to feel there is something special about your community. That you are part of moving toward the future. This population also stands to benefit a great deal from the development of jobs and economic opportunities that offshore wind energy could provide, as well as from the infrastructure like a microgrid to ensure that nearby areas can have reliable electricity when the next floods hit. The indefinable feeling of being involved in something making their community stronger and our world better is challenging to measure. One unfortunate limit of the research covered in this book was not being able to talk to Venice Park residents directly and hear their reactions thoughts on the wind farms. This would be an excellent topic for ongoing research.

Evolution

Throughout this book, the evolution and transformation of how we make and distribute our electricity are covered in various ways. Rick has led the ACUA to fully embrace moving into a clean energy economy, and his insight into this speaks volumes:

In any disruption of an existing system, there are potential losers. You have to start the dialog . . . let's start planning and evolving towards that and making the right, appropriate choices, and designing our system, our training our employees, to make these systems work. But as in anything, there's always a segment of the population that wants the status quo, because they know the status quo. The unknown or the new is scary. And there are reasons why it is scary because they've seen disruption. My father worked in an automobile manufacturing plant in Philadelphia. There were 12,000 people who worked there. I worked there one summer when I was in college in the 1970s. That plant is closed. Not one person works there now. We still have cars. But the industry changed. It wasn't because those people didn't do a good job or didn't work hard, but the economics of the situation.

His perspective speaks to the fact that utilities have, in some places, resisted the integration of renewable energy for varying reasons. However, in South Jersey and throughout all of New Jersey, this is not true. In fact, New Jersey was one of the earliest states to add clean energy in cooperation with their utilities. Rick continued:

Companies that are ahead of that, which retrain their employees, and redirect their initiatives and investments, in a prompt way, can help smooth out the bumps and the impacts of those sorts of changes. But those changes are inevitable.

He also offers a perspective on how energy sources have changed somewhat and that this change and growth is natural and beneficial:

I went to my sister's house over the weekend. There is a door next to her driveway, a steel door, where the coal used to get poured down to the cellar. That house has not used coal, probably since the 30s. It had oil, after coal, and now has natural gas. Maybe they're going to have solar, maybe something else. Maybe they're going to have a battery. So it evolves.

FISHERMEN'S ENERGY

Rick suggested I contact Jeanne Fox, currently a Columbia University and Rutgers University adjunct professor and co-founder of the Center for Renewables Integration. Jeanne was the President of the New Jersey Board of Public Utilities from just after Governor Christine Todd Whitman's tenure, through Governors Jim McGreevey and Jon Corzine. She remained a Commissioner under Chris Christie. She is widely recognized for her public policy development, which led New Jersey to be 2nd in the nation, after California, in installing solar power in 2009. During a phone call, Jeanne insisted that New Jersey would also have built the U.S.'s first offshore wind farm if Governor Chris Christie had not continually denied the project during his run for President.

Governor Whitman, whose term ended in 2001, had put in place initial infrastructure and funding for solar in New Jersey. Starting in 2002, Jeanne Fox took these and ran with them. She listened to the public in developing clean energy policies through public town halls which translated to real-life projects like the ACUA's five wind turbines in 2005. During our exploratory phone call, Jeanne recounted holding standing-room-only public meetings that went until 1 a.m. She held these to understand the way the public felt about clean energy and as a means of education about climate change, solar energy, and wind power. Clean energy is rife with misinformation, stereotypes, and just plain ignorance. In 2002, it had not had the commercial successes it has today to use as evidence to disabuse people of these incorrect stereotypes. Jeanne began looking for ways to engage a wider array of people and small businesses in using clean energy and believed that creating public buy-in would help integrate solar and wind faster. She was right. By 2009, New Jersey had risen to its second-in-the-nation ranking for solar.

In attendance at one of these public meetings was a group of fishermen. They had come to speak out against offshore wind. As an industry that relied on natural resources for a living, the fishermen believed that offshore turbines would ruin marine environments. The concern is widespread, as building the turbines is a sizable undertaking; the noise and construction would undoubtedly involve some disturbances to the ocean. However, research in Europe has shown that the turbines can actually have the exact opposite effect—they function like reefs and become a habitat for fish and marine life.[9] Some researchers describe European wind farms as "de facto marine sanctuaries," because there are limits on fishing, and the underwater foundations have gradually turned into artificial reefs, attracting a variety of fish and dolphins.

As with any new development, results from wind installations that had negative impacts on ocean life should also be used to inform any future offshore wind installations; for example, by planning the timing of construction so that it does not interfere with the migration of whales, turtles, and other species.[10,11] Research should continue to look at the effect of offshore wind farms on marine mammals, shellfish, and ocean habitats to determine how to build turbines so they are beneficial rather than destructive. Research should also continue to look at the impact of fossil fuels on ocean life. Science has shown that climate change is warming the oceans and hurting fisheries and ocean life in well-documented ways. One of these is causing fish populations to migrate northward to find cooler water.[12,13,14]

As they learned more about offshore wind and energy, the fishermen changed their minds. In a turnaround, they formed Fishermen's Energy. First, they did a feasibility study (and had the example of ACUA as a success story) and assembled a world-class team of professionals, including engineers experienced in the design and construction of offshore wind farms. Then, they drafted a proposal to build an offshore wind farm. At the time, the New Jersey Legislature, and then Governor Chris Christie, had signed state bills supporting offshore wind development, including building a small wind farm off of Atlantic City.

Fishermen's Energy was poised to become the first offshore wind farm in the nation. They won a $47 million grant from the Department of Energy. Paul Gallagher was Chief Operations Officer of Fishermen's Energy. He describes how he saw this change of perspective from being opposed to wind energy to supporting it:

> When I was a little boy here in Atlantic City, I used to watch the clam boats leave Gardner's Basin and they would turn right. And today, if you go down there, you watch the clam boats and they turn left because even the clams have migrated due to climate change. So the real response here is that these fishermen whose families have been working in these waters for hundreds of years, who were stewards of the environment, saw an opportunity not only to protect their industry, but to protect our world.

Academic research very clearly supports Paul's statement—that warming oceans and climate change have the potential for severe economic effects on fishing industries and beyond. Fish and shellfish migrating is one issue; another is hurricane damage such as that wrought by Hurricane Sandy. Climate change, sea-level rise, and warming oceans will affect everyone who lives in coastal New Jersey.[15]

> When Governor Corzine was governor of New Jersey, he was very interested in developing offshore wind potential from New Jersey's coast. The fishing industry was initially opposed to it, but after a meeting with Governor Corzine, where he told them "it's a big ocean, tell me where I can't build these things," the fishing industry, behind the leadership of our chairman, Dan Cohen, decided to organize a reaction and a way of participating in this industry. They started a company: Fishermen's Energy. They decided to be agents of change instead of victims of change. These 12, 13 investors who started the company operated about $700 million worth of commercial fishing. They're primarily located in Cape May, New Jersey and also have operations in Atlantic City.

Involving all stakeholders in a region is one of the most effective ways of galvanizing support—by marrying regional and local economic development with renewable energy development. It also helps different business interests avoid mistakes, like siting energy sources in ways that create severe, long-lasting damage. These siting errors happened with onshore wind. Fishermen's Energy was founded, partially, to help guide offshore wind so that its development could avoid damage to ocean ecosystems:

> If they had just stood by idly and the government had decided that "this is a good place to put an offshore wind farm," maybe they would have put it in

a prime scallop field. In fact, that almost happened in Massachusetts, but by organizing the fishing industry, particularly the large commercial fishermen, they're able to draw on our own experience, participate in the stakeholder meetings, and let the government know where the fish are—and where the fish aren't. Where the right places to build wind farms are, and to take Corzine's message literally.

At least initially, stakeholders in New Jersey seemed to align behind a shared goal. What Governor Corzine stated—"tell me where I can't put them"—is what the members of this particular fishing industry expected would happen. They are not recreational or small-scale—these are commercial businesses with fleets of boats that depend on reliable fish and shellfish stocks.

Other issues helped position Atlantic City exceptionally well for this new development. Many people associate Atlantic City with its handful of casinos and the popular boardwalk. With limited other industries or job growth, the city has struggled economically for quite a few years. The existing onshore turbines serve as a way for people to remember Atlantic City outside of the casinos and boardwalk and to switch the narrative from a fading economy to something new. The local support combines with infrastructure and policy architecture to create a very cohesive set of circumstances, according to Paul:

I knew that Atlantic City loved their wind turbines. I knew that the conditions were good here. I knew that there were interconnections.[16] There were a lot of the pieces that you need to make the puzzle work right here in Atlantic City. By the time Governor Christie had arrived, there was legislation in the works, led by Senator Kean and Senator Sweeney.[17] When that got drafted, they put in a section which called specifically for an offshore wind farm in state water in Atlantic City. That was what we had focused our interest on originally. We always thought, "let's take a small step first, build a demonstration project, learn not only for ourselves but for the whole industry and the state, and move forward from that point to the larger projects that the federal government would be bringing online later."

The plans for Fishermen's Energy were at a scale that it could be built quickly and relatively close to shore, positioning this small wind farm to be a valuable source of research on the impact of offshore wind turbines on this stretch of the ocean:

We planned to build six turbines, four megawatts each, 2.8 miles off the coast of the boardwalk. The state line ends at three miles. So it is clearly in state waters. It was permitted by state agencies. We decided that the state was in a position to move faster than the federal government when we started this in 2010, and when we got our permits. We were fully permitted by 2011, and we were ready to proceed to build a project. Then, things changed in Trenton. At some point, we withdrew from the greenhouse gas initiative, solar kind of slowed down, and offshore wind never took off.

When Paul says, "we withdrew," he refers to the Regional Greenhouse Gas Initiative, which New Jersey initially helped found in 2005. Under Governor Christie, New Jersey pulled out of this agreement, although the state has since rejoined. The current state administration's goal is to achieve 100 percent clean energy by 2050.[18]

This interview took place in October 2016. Fishermen's Energy had already applied to the Board of Public Utilities for the state, which is the regulatory body that oversees utilities. Fishermen's Energy had to apply for permission to sell their electricity into the grid as a distributed generation source. Paul continued describing Fishermen's Energy progress toward approval from the BPU, which would allow them to start construction:

We have some ongoing conversations and I'm optimistic that in the near future, we'll sign a power purchase agreement. You know, we'll satisfy the Department of Energy requirement . . . they have 117 boxes on their application. And we have 116 of them checked off. The 117th is "who's going to buy your power." We're close to that. I think we'll get there soon and then we'll be ready to build. And we hope to build this project in 2018 and have it operating before the end of 2018.

A few months later, in early 2017, I met with Paul again for a follow-up interview. Instead of an office full of model wind turbines and schematics, the room was empty. The desks gone, the bright banner taken down. Only a few hardhats remained on a card table in the middle of the room. Fishermen's Energy had shut down as a result of not being able to check that 117th box. Paul explains, alluding to the 2016 federal election as the impasse:

Towards the end of last year, we thought we had an offtake agreement for our power, and I've come to believe that the only thing that really matters is having someone to buy your power. We were very close. In mid-December, we found out that the deal wasn't going to close . . . that the counter party had legitimate concerns about spending more money for energy than energy costs in the marketplace at a time when they thought other core values were going to be impacted by the change in Washington.

So, we ran out of time, and we ran out of money, and that was December 16 when we found out that the contract wasn't going to go to the conclusion. We

> brought the staff in, there were six or seven of us. We let everybody go on December 23rd, and since then, I've been closing down the office and winding down the business.
>
> We still have a fully permitted 24MW wind farm. And the permits are going to be good for years, some for five years and some for 25 years. It could be built tomorrow if somebody came along and wanted to buy the power. We've a couple of leads we're chasing down, nothing looks overwhelmingly promising. We basically have talked to some of the other competitors and asked if they were interested in purchasing the project, but we need to either maintain the asset until the times change, and times will change in New Jersey with a new governor. Or, sell the assets to somebody who's got the patience and the capacity and the capital to carry it for a while until the market changes.

I asked him how he had heard the news:

> [After the meeting where they discussed this deal with the potential purchasers] *I met my wife at St. Patrick's, and we were walking back to our hotel and when we got to 55th street* [in Manhattan, near Trump Towers]. *There was all this commotion, and cops were walking around with bike racks and stuff. They were blocking traffic because the President-Elect* [Trump] *was leaving New York. Traffic was stopped for about 20 minutes. As I was standing there, waiting to cross the street, waiting for this to all end, Chris called me* [Chris Wissemann, the CEO of Fishermen's Energy] *and told me the deal was off. We weren't getting the power purchase agreement. And I watched these 35 cars drive by, and I think, "wow, if the election had turned out differently, this would be a very different moment for me."*

BIRDS AND WIND TURBINES

Much of this book focuses on narratives about clean energy: public perceptions which are deeply embedded regardless of how accurate they are. For wind energy, one enduring narrative that begs a focused section is: "wind turbines kill birds." I was interested in knowing from the experience at ACUA if this narrative was accurate. As someone whose background and training is in the creation of narratives, creating ones that "stick" is a key goal in media. It is how the press cuts through a vast amount of content and forges perceptions, framing, and stories that people cannot forget. When applied to marketing or advertising, "sticky narratives" are very effective— so compelling that sometimes narratives stick even when facts disprove them. There are various ways to make a narrative sticky one is to use fear, as humans retain information about frightening or dangerous things much longer and more vividly than we do about things that are benign or even positive. Another way is to simplify. A third way is to repeat the same narrative repeatedly, whether or not it is correct. This marketing technique even has a name: FUD, or "Fear, Uncertainty, and Doubt."[19] FUD narratives have been used for decades. With a kernel of truth, they create a queasy feeling of uncertainty, a doubt about whether or not something new is valuable, and fear of taking any steps toward some type of progress. I wanted to know if "wind energy kills birds" is a FUD narrative or if was indeed a regular and severe outcome of wind farms. In fact, many studies that I found acknowledged that habitat loss is the biggest killer of birds and other species, not collisions with wind turbines.

This section looks at some reasons this narrative around wind power developed. I also share more resources for ongoing research on wind power and its impact on wildlife. Like many false narratives about renewable energy, there are facts supporting this point, enough for it to have traction. Yet, as with all FUD narratives, the truth of the situation is complex and layered, based on partial information, and not informed by current information.

I turned to the growing body of research on bird kills by wind energy and also by energy production in general. The fact is that all modern energy production systems kill birds. Modern energy production systems kill many other species, as well, including humans. Damage and death to bird populations and habitats are of critical importance; however, they are not something that only wind farms cause. Energy production is a system that begins with the extraction of fossil fuels from the ground, from strip mining of coal to oil drilling, fracking natural gas, mining the metals and minerals that make wind turbines and solar panels. The system continues from the level of extraction through gas pipelines, building power plants, operating those plants, producing toxic waste and carbon emissions, and the long-term impact of waste and emissions. The power production—whether at a wind farm, a solar farm, or a natural gas-burning power plant—is only one step in this long process. Many would argue that the long-term impacts that our energy system has on climate change might be most important: how long the toxic waste remains in the atmosphere, how far it spreads, whether or not it decomposes, and if not, whether it can be disposed of safely.

With this kind of end-to-end system in mind, quite a few extensive academic studies have shown that wind power from all the types of energy production, is actually one of the least destructive types of energy production in terms of its impact on avian life. In fact, The National Audubon Society suggests wind turbines actually protect birds. The Office of Energy Efficiency and Renewable Energy published research showing that "wind projects actually rank near the bottom of the list of developments that negatively impact wildlife and the environment."[20] In 2020, The Audubon Society published a statement stating that *"All forms of energy—including wind power—have impacts on birds*" and that "*while wind energy helps birds on a global scale by curbing climate change, wind power facilities can harm birds through direct collisions with turbines and other structures, including power lines.*"[21]

This second statement, from the Audubon Society, is a far more nuanced and intelligent way to discuss the negative impact of wind turbines on birds than the stickier but skewed and inaccurate narrative of "wind turbines kill birds." As a more accurate narrative, it lends itself better toward a solution than an approach that assumes all wind turbines and all wind farms have the exact same effect on birds. The narrative of "wind energy kills birds" is a very narrow approach and seems to be based on a few specific wind farms built early in onshore wind development. These were sited poorly and constructed in the known flight paths of birds and bats. As early developments, they also did not include any of today's newer technology that helps birds avoid collisions with turbines. This narrow and outdated approach concerned me; as someone who cares deeply about the environment, supporting renewable energy is wanting to support sustainable ecosystems. However, I had to ask whether this outdated narrative might actually be causing damage, as it does little to inform the public about the fast-developing technology of wind energy, nor does it alleviate the conditions that can lead to bird collisions. More to the point, other parts of our energy systems kill birds in larger numbers and have been for quite some time. It also is a narrative based around the realities of onshore wind farms—not offshore ones.

What is true is that some wind farms, most notoriously, one particular wind farm in California, were built in locations with little to no attention to bird migration patterns. In the many years of operation, academic research and environmental policy have acknowledged these mistakes, and newer wind farms are using technologies like different colored blades and slower turning turbines, which cause far fewer bird-turbine collisions. Because of the many collisions with birds, specifically raptors such as golden eagles, the notorious California wind farm turned off its turbines. It has since been re-opened, with newer wind turbines that are supposed to allow the birds to pass through and are built so that the birds can see them more easily. As these less-harmful wind turbines come into use, ongoing studies reveal effective ways to reduce the damage that wind turbines cause to birds and bats, which are also heavily affected by collisions with wind turbines. As researchers and scientists find solutions, energy developments need to continue enacting them to reduce or eliminate the damage to birds, bats, and other species.

How does this relate to FUD narratives? To begin with, the narrative about wind turbines looks at the turbines in isolation, not in comparison to fossil fuel power production. It is safe to assume we will continue to need energy at today's levels. Taking a comprehensive look at the energy industry—all of it—would be a more accurate picture than focusing on one type of power production. I would invite the reader to continue even deeper research and come to their own conclusions about whether what they know about wind power, or any type of renewable energy, is accurate or a narrative based on fear, uncertainty, and doubt?

This chapter is the most detailed discussion of negative narratives about energy in this book. One observation, with respect to this particular issue of killing birds, which is highly emotional and broadly publicized, is that ongoing research can shift academic and public concern from the specificity of killing birds to focusing on how we can develop energy systems with the lowest possible overall destruction of birds, other wildlife, habitats, air, and water. Another observation is that with climate change advancing at an alarming pace, it is far more crucial than ever that students, researchers, and professionals continually find the most reliable information, recognize false and harmful narratives, and instead foster thinking that generates solutions and new designs. We need to move away from simplistic and "sticky" narratives that are not borne out by current facts and are likely due to FUD marketing techniques. In other words, we need to seek and foster transformation in our thinking as well as in our energy systems.

Onshore vs. Offshore Wind

When I told people that one storyline in the documentary was following an offshore wind farm development near Atlantic City, a frequent response was: "doesn't wind power kill birds?" While many people were excited to hear about the possibility of offshore wind, the frequency of this response and the relatively slight amount of other information these same people had about wind power revealed another concern. It is essential to distinguish *onshore* wind farms—in valleys, mountain ranges, and marshes—from *offshore* wind farms. The proposed Fishermen's Energy site described in this chapter would be located two-and-a-half to three miles offshore—close enough for some shorebirds to fly through but certainly not the numbers of birds and bats that would occur onshore. The larger wind farms proposed by the end of filming were going to be located more than 15 miles offshore; I was told by many that there are very few birds at all in this part of the ocean.

However, the ACUA wind farm was in the marshlands surrounding the wastewater treatment plant—close enough to have plenty of shorebirds. When I first interviewed Rick, I asked him about bird deaths. He said a researcher had done a before-and-after study on the number of bird deaths at the ACUA wind farm, which is home to many birds and other species. The assessment found exactly the same number of dead birds *after* the turbines were installed as they had found in their assessment before the turbines were installed. In other words, no more birds perished from ACUA's five turbines than would have otherwise died.

ACUA's wind farm was as close to the shore as possible without being in the water. What does this before-and-after assessment reveal for Fishermen's Energy? One of the most valuable aspects of building a small wind farm is to allow for these types of studies before investing in the expense and years of construction of much more significant energy infrastructure. Part of the plan for Fishermen's Energy was to engage with ocean researchers to study the wind farm's impact on both birds and ocean species: whales, fish, shellfish, and the ecosystem in general.

This body of research on offshore wind in the U.S. is still growing, as there is only one small offshore wind farm in the U.S. However, studies on offshore wind in Europe can reveal more about how offshore wind affects wildlife.

Fossil Fuel Production and its Impact on Wildlife

Any discussion of whether wind energy kills birds would be uninformed if it does not include a discussion of other types of energy production. A growing and substantial body of studies indicate that wind turbines *do not* kill an excessive number of birds, particularly in comparison to nuclear and fossil fuel power production, which studies show as being ten times as lethal to birds as wind energy.[22]

Some arguments against wind power, based on its impact on birds and bats, compare the death rates from bird collision with turbines to those turbines not existing at all. In other words, wind turbines kill more birds and bats than if no wind turbines were built. However, this is not a fair comparison. Energy production is a given: we will continue to use it and need to get it from somewhere. A more fair and accurate comparison would be to contrast the number and type of bird deaths from wind turbines to the number and type of bird, bat, and other casualties that would take place to create the energy we use. In other words, wind turbines cause deaths to birds and bats, but would a wind farm cause more death and damage than a fossil fuel plant producing the same amount of energy? Some studies measure the rate at which birds and bats are killed per megawatt of energy. This type of metric might become increasingly valuable, as would other metrics which allow these studies to portray the environmental impact of an energy system with greater accuracy.

It would be more accurate to compare wind power's impact on birds, bats, mammals, humans, and air and water quality to the same rates for coal mining, drilling for natural gas and oil, the toxic waste produced from all of those, the building and maintaining of pipelines, the damage caused by fracking, the operation of nuclear power plants, and the storage and leakage of radioactive waste. Other metrics look at which birds are affected, with statistics showing that raptors and endangered golden eagles are impacted more profoundly, and their deaths are more detrimental overall than other types of birds because their reproduction rates are much slower. These metrics help understand and prevent collisions. However, they are significant in specific regions and less critical in different areas using wind energy. The reason I believe this to be a FUD narrative is that the idea that the vast oversimplification of any FUD narrative applies the narrative of "turbines kill birds" equally without respect to regional differences, wind farm size, correct siting, attention to migration patterns, and the actual count of bird deaths. It also does not consider that offshore wind farms will have a completely different interaction with birds than onshore turbines.

We cannot look at wind power—or solar or geothermal—as individual procedures not related to anything else. Instead, we have to look at our energy production system and each of these types of production within that system. So, the logic of saying "wind turbines kill birds" completely changes if we ask "are wind turbines killing more birds, and which birds and other species, than other forms of energy production." And, "how can each of these systems evolve to reduce this destruction." We need to look at energy systems, not energy silos.

Electrocution from Power Lines

Turning to current academic research, a sizable amount of research looks at the death toll for birds and other species from all phases of fossil fuel production: mining, drilling, fracking, habitat destruction, pipelines, oil leaks, oil spills, strip mining, global warming, acid rain, mercury poisoning, pollution, and toxic waste. Among these, high transmission wires have been a significant culprit of bird deaths for decades. Since the 1970s, bird and wildlife advocates, utilities, and state and federal authorities have all recognized that birds—and very frequently raptors—die from electrocution when they land on high transmission. This happens so often that high transmission wires are considered responsible for the largest number of energy production–related bird deaths. These high transmission wires have no relationship to the type of production used to create the electrons they carry—they could be carrying electricity produced by coal, natural gas, oil, solar, or wind energy. A 2021 article by the National Wildlife Federation shared a study by the U.S. Fish and Wildlife Service's National Forensics Lab, which looked at the deaths of 417 raptors from 2000 through 2015: 80 percent of these birds were bald eagles or golden eagles; the article speculates that this might be due to the overall growing population of raptors and bald eagles from success in protecting habitats and protections. This study looked at, on average, 27 deaths per year of these raptors by electrocution.[23] This preponderance of bald and golden eagles is very sympathetic—they are majestic, beautiful creatures. They also are dwindling in population. Of course, their deaths should be prevented and reduced whenever possible.

However, even more telling are two other facts: one, from a 2014 study which combined data from 14 different studies to "estimate that between 12 and 64 million birds are killed each year at U.S. power lines, with between 8 and 57 million birds killed by collision and between 0.9 and 11.6 million birds killed by electrocution." A spread of 0.9 million to 11 million bird deaths is a vast range, particularly when compared to 27 deaths per year on average. It also indicates that the energy transmission systems are as much of a killer as are the energy production systems.

More alarming than the possibility of 11 million bird deaths or a growing number of eagle deaths, is that the same 2016 U.S. FWS study stated that "the incidence of electrocution does not appear to have decreased despite over three decades of research and mitigation procedures," a finding that is even more significant than the damage done by wind turbines. Moreover, it places accountability on human negligence in responding to these deaths, as well as on energy infrastructure.[24]

Regardless of the type of electrical power production, if the goal is to prevent birds from dying, especially endangered ones like raptors, it seems like a better solution is to create "avian-friendly transmission lines" rather than stopping the development of wind farms. The American Eagle Foundation has proposed this type of transmission line.[25] As mentioned above, there are also wind turbines that have been developed and are being used that are less deadly to birds, including raptors. Studies from these turbines show significantly less damage and death to birds when using these newer turbines. Compare this to the FWS study that states that deaths from electrocution have not gone down, and it seems like the facts bear out that one industry is responding while another side is not or cannot.

Additional research has also shown birds perish by colliding with smokestacks, in oil pits and evaporation ponds, by crashing into communications towers, and flying into windows in far greater numbers than they do from wind turbines.[26] A few studies have aimed to quantify this. For example, a 2009 study compared bird deaths by wind turbines to bird deaths from nuclear- and fossil fuel-burning power plants, finding: that wind farm-related avian fatalities equated to approximately 46,000 birds in the United States in 2009, but nuclear power plants killed about 460,000 and fossil-fueled power plants 24 million.[27] A comprehensive Canadian study from 2013 dug further into avian deaths and looked at selected causes of bird death. This study looked at an annual total of 186,429,553 estimated bird deaths caused by human activity, with the cause of death being a variety of factors, including energy systems. They found: "combined, cat predation and collisions with windows, vehicles, and transmission lines caused greater than 95 percent of all mortality."[28] Most recently, an Australian researcher reviewed several studies, including these two as well as others from Spain and Australia, concluding that wind farms "are hardly the bird slayers they're made out to be." He noted that the British Royal Society for the Protection of Birds "built a wind turbine at its Bedfordshire headquarters to reduce its carbon emissions (and in doing so, aims to minimize species loss due to climate change)."[29]

The facts are not meant to minimize the importance of reducing collisions between birds and wind turbines. Instead, they are included here to separate the quality information from the FUD narratives. Continuing to research harm to birds, marine mammals, and ecosystems caused by our energy production system is very important, as is continuing to deepen our understanding of how to minimize this damage as much as possible. However, ever-growing bodies of research show that fossil fuel energy production, habitat loss, and transmission lines kill far more birds than renewable energy. And that the beneficial aspects of clean energy—the capacity to mitigate climate change, reduce pollution and toxic waste, and create local production and reduce the need for high voltage transmission—far outweigh its dangers.

Time for Turbines

Fishermen's Energy did not get the buyer for their power, but Paul did get one of his wishes: a new governor. In 2018, Governor Phil Murphy took office, running on a progressive slate that included wind power development. It seemed there was every intention of finally realizing the potential to create a local wind industry. With the new government in Trenton, Fishermen's Energy, the ACUA, and many other stakeholders drew together. They held a conference, Time for Turbines, which brought together people working in wind power, labor, environment, and policy. It took place at the ACUA and was hosted by the Business Network for Offshore Wind and New Jersey Renews and also attended by legislators who governed renewable energy policies.

In 2018, the mood at Time for Turbines was highly optimistic. One of the legislators said at the mic, emphatically, "New Jersey is open for business regarding offshore wind!" Ørsted, a Danish company and the global leader in offshore wind, announced that they would open an Atlantic City office the following month.

With a new governor in office, there was every reason to believe that somebody would put "steel in the water," possibly even that summer or fall. Fishermen's Energy had always envisioned itself as a way to start laying the groundwork for much bigger wind farms. As a pilot-scale project, it could be a site to study how to best build wind turbine foundations with minimal effects to the ocean floor, marine species, birds, and fishing industries in New Jersey. While existing research covered many of these, much of the research was done on computer models or in other countries or states simply because there were no actual wind farms to study in New Jersey. They also saw a role in starting to develop onshore resources and infrastructure, training personnel, and otherwise beginning the development of the local wind economy as a kind of test. Companies like Ørsted build wind farms with hundreds of turbines, and there seemed to be value in a smaller project that was completely permitted and ready to go, which could also serve to work out many remaining unknowns about the industry.

Fishermen's Energy had found an investor in EDF RE, a French offshore wind company, and had decided to reapply to the Board of Public Utilities. A news report noted their expectations to "improve environmental management by providing a laboratory for testing of new avian monitoring and marine mammal sensing technologies."[30] The third time might prove the charm. Fishermen's Energy filed again with the BPU to gain approval for their project, this time with a major international wind company on their side.

Eight months later, it came to an end: for the third time, the state of New Jersey rejected Fishermen's Energy's plan. The decision, in late December 2018, was met with disappointment. The President and CEO of the Business Network for Offshore Wind, Liz Burdock, spoke to the value of Fishermen's Energy beyond just clean power, as it would have provided an opportunity to continue developing their supply chain and the business community for wind.

The BPU President Joseph Fiordaliso stated the reason for rejecting Fishermen's Energy was the price of its electricity, which would cost more than other projects he had seen. In news reports, he did not reveal the price detailed on the application, citing confidentiality, although he did cite New Jersey's wind development law, which requires developers to show a net economic benefit for ratepayers who pay for the electricity.

Offshore Wind Rates and Ongoing Questions

Rates and the consumer's burden of cost are very important in renewable energy integration everywhere, and they are discussed in every chapter of this book in one form or another. As states everywhere start to develop their master plans for offshore wind, is it important to engage the public in knowing what that net economic benefit offshore wind power will have for consumers? And for local governments, utilities, and energy developers to foster public knowledge of how offshore wind farms affect rates? Engaging the public through the very relatable mechanism of the price on their electric bill seems an excellent entrée to fostering deeper public education about renewable energy. It provides a means to help the public see how energy integrates into the grid and how it can foster small business development as well as attract large business investment. It is also a way to show the public the tangible and immediate benefits to a family's budget. For example, in chapter 5, I discuss a Vermont town where an onshore wind farm was developed. The local government received revenue from the wind farm and passed this income on to residents by lowering property taxes. This town voted twice on whether they should have the wind farm: the first time, it passed with a 75 percent vote. The second time, it passed with an even higher proportion of the vote. It seems that people understood how the wind project was designed to create immediate economic benefit for all stakeholders.

Following this story in Atlantic City, while I did not capture the opening of an offshore wind farm with my camera, I did witness many aspects of the beginnings of an industry. From the ACUA's success in using varied types of clean energy technologies to power their 24/7 industrial site, to the years prior during which Jeanne Fox led public town halls that helped integrate solar widely, to the conferences that brought together wind power stakeholders to build working relationships, and the three tries and ultimate rejection of Fishermen's Energy. Combined, these show how complex it can be to get started in clean energy, the type of business ecosystem needed, as well as how a well-planned business can work perfectly.

Early in filming, Paul had said Fishermen's Energy had a goal *"to be agents of change instead of victims of change."* One of the themes of this book is how communities throughout the U.S., sharing this sentiment, found ways to also be agents of change for clean energy. Given that this desire to be part of this change, I am left wondering if this same kind of approach that the Vermont onshore wind project took could be applied to offshore wind development? The laws governing oceans are not the same as laws governing land; however, if the net economic benefit is the issue, this ending leads to a few questions: Can a town or city lease the waters off their shore and receive revenue, as they do with onshore wind and some solar developments? Can business models offer ways to share economic gain with those who live closest to wind farms? Given offshore wind's size, expense, and potential to affect marine ecologies, what is the value of a pilot project for offshore wind development? Is there an economic value to developing infrastructure, a skilled labor pool, ways to protect marine and avian species, and otherwise prepare for much more extensive offshore wind developments? Finally, how can industries that already utilize the ocean work with the burgeoning offshore wind industry? In many places, it is the fishing industry trying to stop the development of offshore wind; in Atlantic City, they were the proponents. In a final observation from Paul, he notes how this shifts the relationship to the economic aspects of wind development: *"The fishermen wanted to be on the inside of the industry. If they can't protect their interests from the inside, then they have to protect them from the outside, and they've been doing this by protesting and objecting in some other places."*

CONCLUSION—BIG WIND, MEGA-REGION

During our initial interview, Paul told me that New Jersey and the Atlantic City coast were considered ideal parts of the Atlantic for offshore wind. The ocean floor is very shallow, with a long sloping bottom and consistent, predictable winds. The Fishermen's Energy founders were not the only ones who saw this potential. New Jersey moved ahead on its offshore wind plan, and in 2019, the BPU awarded Ørsted approval to build a 99 turbine wind farm anticipated to go in the water in 2024. It is planned to have turbines 50 stories tall and will be sited 10 to 20 miles out in the ocean, much further than where Fishermen's Energy would have been.

States up and down the Atlantic coast, along with coastal states around the United States, are preparing their master plans for offshore wind. Other states have already awarded leases to large wind companies for exploration and planning. In New Jersey, Governor Murphy set a goal to achieve 7,500 megawatts of offshore wind by 2035. The 99 turbine Ørsted project, "Ocean Wind," is anticipated to produce 1,100 megawatts, enough power for approximately 500,000 New Jersey homes.

In December 2020, two more offshore wind projects submitted bids to the state: one from EDF Renewables (the former business partner of Fishermen's Energy) and Shell New Energies for a 2,300-megawatt project. The other is a new project from Ørsted, although no details are public about that one at the moment.

These large offshore wind projects, if and when they go in the water, will be right smack in the middle of the largest collection of humans on the planet. The mid-Atlantic is the world's largest "mega-region"—drawing a line from Boston through Connecticut, New Jersey, New York, Maryland, and into Washington D.C., this region is home to almost 50 million energy-hungry people.[31] All of these states have access to the vast potential for energy from the Atlantic winds, and some states are taking solid steps toward accessing that energy. Some of the research estimated staggering potential—one study gauged that New Jersey could supply 92 percent of its current electrical generation with offshore wind,[32] allowing this small and very densely populated state to meet its 100 percent renewable energy goal readily. If New Jersey can create nearly all its power with offshore wind, can the other states in this mega-region also do so?

With one big wind contract awarded and two more in the works, we will continue to watch and wait to see who gets steel in the water next. Although they are no longer competing to be second in the United States: in summer 2020, Ørsted started construction on an offshore wind farm in Virginia, the second in the United States. Who will be third?

NOTES

1. Dominion Energy. "Coast Virginia Offshore Wind: Project Timeline." *Dominion Energy website*. Accessed June 25, 2021. https://coastalvawind.com/about-offshore-wind/timeline.aspx

2. Schlossberg, Tatiana. "America's First Offshore Wind Farm Spins to Life." *New York Times*, December 14, 2016. https://www.nytimes.com/2016/12/14/science/wind-power-block-island.html

3. Firestone, Jeremy, Cristina L. Archer, Meryl P. Gardner, John A. Madsen, Ajay K. Prasad, and Dana E. Verone. "Opinion: The Time Has Come for Offshore Wind Power in the United States." *Proceedings of the National Academy of Sciences of the United States of America*, published online September 29, 2015. doi: 10.1073/pnas.1515376112

4. Jacobson, Mark Z., Cristina L. Archer, and Willett Kempton. "Taming Hurricanes with Arrays of Offshore Wind Turbines." *Nature Climate Change* 4, no. 3 (2014), 195–200.

5. Atlantic County Utility Authority. "Jersey-Atlantic Wind Farm." *ACUA.com, accessed* December 29, 2020. http://www.acua.com/green-initiatives/renewable-energy/windfarm/

6. U.S. Census, 2019 data, 08401 zip code. Data.census.gov

7. Lewis, Andrew S. "Why Atlantic City's Minority Neighborhoods Are Also Its Most Flooded." *NJ Spotlight News*. April 5, 2021. https://www.njspotlight.com/2021/04/redlining-atlantic-city-nj-overlooked-underfunded-minority-neighborhoods-back-bay-racist-maps-superstorm-sandy/

8. National Oceanic and Atmospheric Administration. "The State of High Tide Flooding and Annual Outlook." *Tides & Currents website*. Accessed July 1, 2021. https://tidesandcurrents.noaa.gov/HighTideFlooding_AnnualOutlook.html

9. Keegan, James. "Offshore Windmill's Impact on the Marine Environment." *University of Miami, Shark Research blog*, March 4, 2015. https://sharkresearch.rsmas.miami.edu/offshore-windmills-impact-on-the-marine-environment/

10. Akademie, D.W. "How Do Offshore Wind Farms Affect Ocean Ecosystems?" *DW.com*, accessed September 20, 2020. https://www.dw.com/en/how-do-offshore-wind-farms-affect-ocean-ecosystems/a-40969339.

11. Broom, Douglas. "Reef Cubes': Could These Plastic-free Blocks Help Save the Ocean?" *World Economic Forum*, July 3, 2020 https://www.weforum.org/agenda/2020/07/reef-cubes-arc-marine-biodiversity-wind-farms/

12. Free, Christopher M., James T. Thorson, Malin L. Pinsky, Kiva L. Oken, John Wiedenmann, and Olaf P. Jensen. "Impacts of Historical Warming on Marine Fisheries Production." *Science* 363, no. 6430 (2019), 979–983. doi: 10.1126/science.aau1758

13. NOAA Fisheries Public Affairs. "New Study: Climate Change to Shift Many Fish Species North." NOAA. Last modified 2019. https://www.fisheries.noaa.gov/feature-story/new-study-climate-change-shift-many-fish-species-north

14. New Jersey Conservation Foundation. "Fish In Hot Water." *NJConservation.org*, July 3, 2019. https://www.njconservation.org/fish-in-hot-water/

15. Leichenko, Robin M., Melanie Hughes Mcdermott, Ekaterina Bezborodko, Michael Brady, and Erik Namendorf. "Economic Vulnerability to Climate Change in Coastal New Jersey: A Stakeholder-Based Assessment." *Journal of Extreme Events*, June 2014. doi: 10.1142/S2345737614500031

16. U.S. Department of Energy. "Learn More About Interconnections." *Energy.gov*, accessed December 29, 2020. https://www.energy.gov/oe/services/electricity-policy-coordination-and-implementation/transmission-planning/recovery-act-0

17. New Jersey State Legislature. "The Offshore Wind Economic Development Act (S-2036), otherwise known as the Sweeney/Kean Bill." *NJsendems.org*, accessed December 29, 2020. https://www.njsendems.org/sweeneykean-bill-to-spur-offshore-wind-energy-released-by-senate-committee

18. New Jersey State, "The Regional Greenhouse Gas Initiative in New Jersey." *Department of Environmental Protection*, accessed December 29, 2020. https://www.state.nj.us/dep/aqes/rggi.html

19. Harris, Rhonda. "Chapter 5, Selling Outlook. FUD—Fear, Uncertainty, and Doubt: Those Subtle Messages about Competitors." *The Complete Sales Letter Book*, Armonk: Sharpe Professional. 1998.

20. U.S. Department of Energy. "Wildlife Impacts of Wind Energy." U.S. Department of Energy, Wind Energy Technologies Office. Accessed May 15, 2021. https://windexchange.energy.gov/projects/wildlife#:~:text=Research%20shows%20that%20wind%20projects%20actually%20rank%20near,posed%20to%20birds%20and%20people%20by%20climate%20change.

21. National Audubon Society. "Wind Power and Birds: Properly Sited Wind Power Can Help Protect Birds from Climate Change." *National Audubon Society* website, July 21, 2020. https://www.audubon.org/news/wind-power-and-birds

22. Sovacool, Benjamin K. "Contextualizing Avian Mortality: A Preliminary Appraisal of Bird and Bat Fatalities from Wind, Fossil-fuel, and Nuclear Electricity," *Energy Policy* 37, no. 6 (2009), 2241–2248. https://ideas.repec.org/a/eee/enepol/v37y2009i6p2241-2248.html

23. Greco, JoAnn. "A Shocking Toll: Saving eagles from the lethal hazards of power line electrocution." *National Wildlife*, National Wildlife Foundation. Feb. 1, 2021. https://www.nwf.org/Home/Magazines/National-Wildlife/2021/Feb-Mar/Animals/Eagles-and-Powerlines

24. Lehman, R.N., P.L. Kennedy, and J.A. Savidge. "The State of the Art in Raptor Electrocution Research: A Global Review." *Biological Conservation* 136, no. 2 (2007), 159–174

25. American Eagle Foundation. "Promote Avian-Friendly Power Lines: Millions of Birds Fatally Collide with and Are Electrocuted by Power Lines Annually. There Are Ways to Prevent This!" *Eagles.org*. Accessed May 15, 2021. https://www.eagles.org/take-action/avian-friendly-power-lines/

26. Richardson, Jake. "Wind Power Results In Very Few Bird Deaths Overall." *Clean Technica,* February 21, 2018. https://cleantechnica.com/2018/02/21/wind-power-results-bird-deaths-overall/

27. Sovacool, Benjamin K. "Contextualizing Avian Mortality: A Preliminary Appraisal of Bird and Bat Fatalities from Wind, Fossil-fuel, and Nuclear Electricity." Energy Policy 37, no. 6 (2009), 2241–2248. https://ideas.repec.org/a/eee/enepol/v37y2009i6p2241-2248.html

28. Calvert, A. M., C. A. Bishop, R. D. Elliot, E. A. Krebs, T. M. Kydd, C. S. Machtans, and G. J. Robertson. 2013. "A Synthesis of Human-related Avian Mortality in Canada." *Avian Conservation and Ecology* 8, no. 2 (2013), 11. doi: 10.5751/ACE-00581-080211

29. Chapman, Simon. "Wind farms are Hardly the Bird Slayers They're Made Out To Be. Here's Why." *The Conversation,* June 16, 2017. https://theconversation.com/wind-farms-are-hardly-the-bird-slayers-theyre-made-out-to-be-heres-why-79567

30. Renewables Now. "EDF RE to acquire Fishermen's 24-MW Wind Demo Off Atlantic City." *RenewablesNow.com*, April 5, 2018. https://renewablesnow.com/news/edf-re-to-acquire-fishermens-24-mw-wind-demo-off-atlantic-city-607792/

31. Florida, Richard. "The Real Powerhouses That Drive the World's Economy." *Citylab/Bloomberg.com,* February 28, 2019. https://www.bloomberg.com/news/articles/2019-02-28/mapping-the-mega-regions-powering-the-world-s-economy

32. Oceana. "Offshore Wind Report: Key Findings." *Usa.Oceana.Org.* Accessed January 2. https://usa.oceana.org/offshore-wind-report-key-findings.

Chapter 4

New York, NY
Green New Deal

INTRODUCTION

New York City's famous, glittering skyline is slowly changing. With a population of 8.3 million residents,[1] New York is, by far, America's largest city. That famous Manhattan skyline is the center point of a much broader metropolitan area spanning the five New York boroughs, New Jersey, the surrounding Westchester and Long Island suburbs, and into Connecticut and Pennsylvania. With a population of around 20 million in this metropolitan area, the New York sprawl ranks as one of the top 10 "megacities" in the world.[2]

New York is also old. Making up that famous skyline is a collection of more than 1 million buildings. Looking at that beautiful skyline, they seem like a single, united body. When looked at individually, each of those buildings has a distinct role in New York's energy transition—some are leading the way, and others exemplifying the problems.

Many of those buildings, if not most, have basic infrastructure that dates back decades. Systems like heating, lighting, electricity, and ventilation can be complicated to change in large buildings. Old buildings can have poor insulation, old boilers, and drafty windows that are inefficient and wasteful. Changing these dated systems is a massive undertaking. While retrofits and renovations are possible, ripping out the guts of a building in which hundreds of people live or work is expensive and can take months or even years to complete. Changing these dated systems is also key toward efforts to lower New York City's carbon footprint, as updated infrastructure produces less pollution and uses less energy.

Concurrently, New York City residents, businesses, and city entities are adding sources of clean production of electrical power like solar, wind, and geothermal. Both New York City and the state have sizable clean energy industries. Both have made 100 percent renewable energy commitments, including addressing environmental justice and equity as part of clean energy integration.

Changing how this mass of concrete is heated, cooled, and powered has high stakes for city residents. Pollution has been causing respiratory issues for decades, with rampant asthma, various types of cancers, and other well-documented problems of living with heavy air pollution. COVID 19 compounded existing air quality issues, with many survivors having ongoing lung issues. New York City already had some of the nation's highest rates of hospitalization and death from asthma, with areas in western Queens, northern Manhattan, and the southern Bronx being particularly impacted. The next few years will show if asthma and respiratory disease worsened due to coronavirus.

There are other well-known stakes—in 2012, Hurricane Sandy exposed the city's vulnerability to climate change, rising sea level, and crippling storm surges. It also showed how unguarded our electrical system is to the impacts of climate change. Hurricane Sandy caused blackouts lasting for over a month in some areas, knocking out power for millions. These neighborhoods were left with no lights, power, or heat in a freezing November.

Can clean energy and building efficiency fix these problems? With enough solar panels and wind turbines, can we ensure kids all over the New York metro area grow up free of asthma and that their neighborhoods can survive storm surges without massive power failures? If New York City models green building and clean energy infrastructure, how will it engage people who are often left out of energy decisions, like communities of color, immigrants, and low-income families? How will it create an important precedent and model for megacities everywhere? How will America's biggest city meet its ambitious goals to lower its carbon footprint and use renewable energy?

CHAPTER SUMMARY

This chapter looks at how initiatives such as building efficiency regulations and community solar have an outsize impact in New York, simply due to the city's size and large population. In this vein, this chapter discusses city legislation that has been called the strongest environmental regulations of any city in the world. It also discusses the many bills that lead up to the complex way of clean energy development proceeds in America's biggest city, including a 2021 law that will turn an infamous prison island into a cutting-edge solar farm with resilient backup systems.

This chapter also looks at community activism in Astoria, a neighborhood in the western part of Queens, which helped spur these policy issues. In the early 2000s, Astoria residents stopped the utility from building another dirty, coal-burning power plant in an area with several existing power plants. That spirit of fighting dirty power was carried all the way to City Hall when Astoria residents elected Costa Constantinides as their City Council member. Costa passed dozens of successful bills with aggressive environmental and sustainability goals during his tenure at City Council, impacting all of New York City and with the potential for national influence.

Finally, this chapter looks at the start of New York City community solar programs allowing renters and apartment-dwellers to purchase electricity from a solar farm on the roof of a warehouse, garage, or another industrial facility. These types of community solar projects are only just beginning to emerge in the United States in urban areas. They allow warehouse owners to earn revenue by leasing previously empty roofs for solar development. The electricity from these farms is sold to consumers through the regular grid at a price comparable to or lower than standard electric rates.

METHODOLOGY

My interviews for the film were with New York City Council Member Costa Constantinides and Astoria resident Anthony "Tony" Gigantiello. I also spoke with Professor Rebecca Bratspies, one of the many proponents of the Renewable Rikers plan as well as a law professor focused on environmental justice. I interviewed two companies with community solar—The Power Market and OnForce Solar—both selling solar power from rooftops in Queens, the Bronx, Brooklyn, Yonkers, and many other areas.

I followed Costa's work from 2017 through spring 2020, when Renewable Rikers was introduced to the New York City Council, and then followed the progress of the bill until it was passed into law in spring 2021. We did several interviews, and I filmed at press conferences and town halls, including a press conference to announce the City Hall bill for Renewable Rikers. I also followed City Council meetings and minutes and spoke with Costa's staff to get updates and context for the bills. We wrapped all production at the end of 2019, right before Renewable Rikers was introduced to New York City Council in January 2020. By the time you read this book, hopefully, all of these pieces of legislation will have manifested into tangible projects that generate clean energy. I encourage readers to do their own investigations to learn about developments after 2021.

POWER HUNGRY DEMAND

Bit by bit, New York City's buildings are changing. Many building owners spend the time and money updating their heating systems and making their building's operations cleaner and more efficient. Some are even adding solar rooftops or solar power systems on expansive warehouse roofs. Other buildings are adding small rooftop wind turbines, which look more like barrels, and installing other technology that allows buildings to burn less fuel and create less pollution. In a tiny precursor of what is expected in the next few years, New York City's first and only large wind turbine—the standard type of wind turbine with three large blades—rises from the site of a recycling facility located in the Sunset Park neighborhood of Brooklyn. From certain vantage points, it blends in as part of the harbor view as just another part of this ever-changing megacity. Before discussing Astoria's activism, this section looks on a macro level, at the electrical demand and generation sources of electricity in New York and how both are constantly changing.

On a large scale, initiatives for a significant amount of clean energy in New York are on the horizon, although it could be many years before these finalize and replace fossil fuel plants. One of these initiatives—large amounts of offshore wind—is currently in a phase of New York State requesting proposals from companies to build these wind farms. At the time of writing, there

Figure 4.1   Costa Constantinides and Hundreds of New York Environmental Activists at the Announcement of the Green New Deal. *Source*: La Rosa, Melanie, 2018.

are several offshore wind proposals in development with New York State, which would contribute to the electrical power generation for those 20 million people and millions of others across New York State. These initiatives are meaningful and hold great promise for clean air, affordable electrical generation, and even making New York City more resilient. But, how can we ensure these promises are realized? How can regular, everyday people participate in this growing clean energy economy if someone does not have a warehouse roof to rent or own their home and can install solar panels? While a tiny precursor of wind power is hope-inspiring, with that, along with some solar rooftops, be enough?

One of the most universal challenges of transforming to a clean energy economy is meeting the consistent, round-the-clock demand for electricity. (See chapter 10 about Las Vegas for discussion about the relationship between the sun's output and consumer demand). Without a plan on how to meet the electrical demand of those 20 million-plus people, we cannot merely replace fossil fuel-burning power production with clean energy plants.

From 2010 to 2019, New York City's population grew,[3] adding to an electrical demand. Power plants are located throughout New York City. However, the vast majority are in Queens, with a large cluster in Astoria. It will come as no surprise that while most of the city's electricity is generated in Queens, the vast majority of it is used in Manhattan.[4]

Another critical factor is that city, and state laws require that New York City generate over 80 percent of its "peak load"—the highest power demand on a system over a specific period of time—by power plants within the city limits.[5,6] This differs from many cities, which locate power plants outside the city limits. It also results in health issues like asthma alleys and other illnesses from people living too close to pollution from fossil fuels.

To understand why New York has this requirement, and to learn its practical impact, I asked the founder of the Sabin Center for Climate Change Law at Columbia Law School, who provided a telling answer relating to blackouts:

> This was imposed years ago by two entities—the New York State Reliability Council and the New York Independent System Operator—that have the responsibility to help ensure the reliability of NYC's electricity supply. The reason for this requirement is that there is a limited amount of electricity transmission capacity (power lines) running into NYC. If NYC were too dependent on these lines, and they went down or were otherwise disrupted, there could be a blackout. Hence the requirement that the great bulk of the power used in NYC, be generated in NYC.[7]

There are also many large power plants just outside of the New York City limits and many clean energy developments proposed in surrounding areas.

When asked about the potential impact this 80 percent law might have on renewable energy development, he explained:

> There is work on building special lines that would take power directly into NYC from more distant sources—offshore wind from Long Island, hydropower from Quebec—and since those lines would plug directly into the NYC grid, I think they would be exempt from the 80% limit. The 80% rule is why there are so many power plants in NYC . . . [it] is very good for in-city renewables and very tough for out-of-city renewables.[8]

The regulatory organizations he mentions the New York State Independent System Operator (NYISO) and the New York State Reliability Council (NYSRC)—are the state bodies that manage New York's electric grid and ensure reliable and consistent electricity. ISOs are part of the energy administration in many states, with coordinating generation to meet demand. These organizations also manage input to the system from solar, wind, and other distributed, intermittent power sources to balance fossil fuel power generation.[9,10]

His response indicates yet another reason why the stakes are very high for New York City to find practical, affordable, and reliable ways to make solar power more available and affordable for New Yorkers. Installing a rooftop solar farm is vastly quicker than building an offshore wind farm, and it creates a power source close to where that power is used, which minimizes the need for more power lines. On the other hand, offshore wind has the capacity to produce enough power to effectively replace fossil fuel-burning power production. Combined, they make most consistently at midday (solar) and at night (wind power) and are well matched to both contribute to our grid.

Impact of Coronavirus

The coronavirus epidemic impacted electrical use in the metropolitan area significantly. The NYISO, in early 2021, reported that during April and May 2020, electrical demand was down by a tremendous 8 percent from the forecast developed a year earlier. While usage eventually rose back up, it was not until January 2021 that it reached previously predicted levels. This report also noted that the location of power use shifted from New York City to suburban areas in Long Island and the Hudson Valley, as people worked from home, offices shut down, and commercial and entertainment activity ceased.[11]

This is a shift in daytime electrical use from the standard duck curve discussed in chapter 2, which had lower use at midday then higher demand from 6 p.m. to midnight. This examination of demand, the time of day, and the overall amount of power used provide valuable insight as New York plans clean energy use. It is an excellent area for ongoing research. If the shifts in the time and location of electrical demand seem like they will be permanent, it could influence how New York City plans clean energy integration.

Nuclear Power—Indian Point Shuts Down

In a significant victory for environmental health, the state legislature voted to close a nuclear power plant about 20 miles north of New York City, Indian Point. This law was passed overwhelmingly in both houses of the New York State Legislature. In April 2020, one of the two reactors at Indian Point was turned off for good. The other was shut down in April 2021.[12] These two units began operations in 1974 and 1976 and were well past their expiration dates. One of four nuclear power plants in New York, Indian Point made 12 percent of the total electric mix for the state.[13]

Indian Point shutting down was a milestone for environmentalists who had long opposed this aging nuclear power plant still operating just upstream from New York City and its sprawl of people. Unfortunately, New York City is striving to install solar panels and wind turbines in sufficient numbers. When Indian Point shut down, three natural gas power plants were brought online in upstate New York and New Jersey to replace the power from Indian Point.[14] This somewhat mitigates the environmental health success, as essentially, New Yorkers have traded the specter of radioactive waste for methane gas, fracking in multiple states, and incentive to build more highly controversial pipelines.

These natural gas plants are supposed to be temporary, until New York builds the expected offshore wind projects that are currently under development. The first one of those is anticipated for 2023.[15] At a time when renewables are within reach, and some cities are hitting historically high rates for the proportion of their energy mix from solar, wind, and other clean energy, it is worth asking why the city and state did not move faster to approve the offshore wind farms which have the capacity to replace Indian Point's power levels? An even better question: With temporary fossil fuel power production replacing Indian Point, how will the city and state speed up the process to dramatically expand the use of solar and wind?

This macro-look at how one source of electrical generation can shift into another source is meant to provide a birds-eye view of our current energy system. It shows that it is also constantly changing. For a different, hyperlocal look at how residents are actively changing power systems, we go to Astoria, Queens, where community members shut down a coal-burning power plant and are pushing for renewable energy in a historic city law called "Renewable Rikers."

Chapter 4

ASTORIA WINS—FROM THE POLETTI POWER PLANT TO RENEWABLE RIKERS

In the early 2000s, the New York utility—ConEdison—planned to build yet another coal-burning power plant in Astoria, Queens. Astoria, located in the northwest corner of Queens, is across the East River from Harlem, with the Bronx to the north. Right in between them is Rikers Island, located where the East River flows into the Long Island Sound. Rikers Island is accessible from only one bridge, which leads to Astoria.

This is important for various reasons: Rikers is located in a very inaccessible part of the city, only reachable by infrequent buses or expensive cab rides. This is only one of the many reasons that the prison on Rikers Island has been planned to close in 2026.[16] It is located right between two environmental justice areas, which suffer the consequences of infrastructure in their neighborhoods, primarily power plants, transit and shipping, and wastewater treatment centers. There are other reasons to shut down Rikers Island. Rebecca Bratspies, a professor at CUNY School of Law and Director of the Center for Urban Environmental Reform, put it this way: "In every way, Rikers is toxic."[17]

Rebecca lives in the same complex of buildings as Tony Gigantiello and was also heavily involved in the protest and shut down of the Poletti Power Plant. Her account of Rikers history speaks volumes about the need to reclaim it for environmental justice purposes. In short: Rikers Island, originally the land of several tribal societies including the Munsee Lenape, Wappinger, Matinecock, and Lekawe/Rockaway Tribe, was granted to the Riker family by the Dutch government when the area that became New York City was still a colony. The Rikers were a slave-owning Dutch family who built their fortunes through the labor of enslaved people; one of their most infamous family members was "Recorder Riker," a city official who ran the court system. He was notorious for declaring free black New Yorkers as "fugitive slaves" and selling them into slavery. Originally, the tiny island just off Astoria's northern coastline was under 100 acres. New York City bought it to use as a jail in the late 1800s and used inmates to shovel Manhattan's waste and expand the island to four times its original size. From every perspective, the history of this island, made out of landfill and used for centuries as a jail, begs redemption.

That redemption could be a renewal as a site for environmental justice, green jobs, and healthy infrastructure. A tiny island made out of landfill is not suitable for most uses, but it is ideal for a solar-powered battery storage system and wastewater facilities. It is virtually treeless and set in the middle of the East River, with constant and unobstructed sun. As residential batteries start to come into consumer use, utilities are also testing utility-sized batteries and their safety parameters. A small island physically separated from both Queens and the Bronx is the perfect place for testing utility batteries in a dense urban setting. The sizing of the system would allow it to create enough clean energy to shut down all of New York's "peaker plants," which generate a lot of pollution (see Renewable Rikers section below for more about peaker plants). Renewable Rikers also proposes to relocate pungent wastewater treatment plants out of Queens and the Bronx to Rikers Island, moving the putrid smells away from neighborhoods. (In chapter 3, a discussion of how a wastewater treatment center in Atlantic City became a 100percent renewable energy site speaks to how this is possible.)

In January 2021, the New York City Council passed a law that would turn Rikers Island into a solar farm when the infamous prison is closed. (See section below for more on Renewable Rikers plan.) This plan was years in the making and involved many people in the environmental justice and equity fields. The bill was introduced to the City Council by Council member Costa Constantinides, and the roots in Astoria are deep.

The Poletti Power Plant

In the early 2000s, Astoria residents were coping with high rates of asthma, respiratory disease, and throat cancers. One of the many power plants located in the area—the Charles Poletti Power Plant—alone was responsible for more toxic emissions than all of the power plants in neighboring Brooklyn, as reported by media outlets from Queens local papers[18] to the New York Times.[19]

Tony Gigantiello, a local resident and community activist, had been president of the board of his co-op apartment building for many years, a complex with over 300 apartments. His co-op building is located less than 3 miles from a cluster of five power plants and a wastewater treatment center, typically collectively referred to as "Astoria Energy" (although it is really several different companies on one site). This complex takes up the entire northwest corner of Queens, is fronted by the East River waterfront on one side, and 20th Avenue on the other.

On the other side of Tony's co-op building, less than a mile away, is the Ravenswood Power Plant. The 4-mile stretch from Astoria Energy to the Ravenswood Power Plant is a residential urban neighborhood packed with homes, co-op apartments, public housing, parks, schools, and small businesses. Ravenswood Houses, a public housing complex with more than 2,000 apartments, is one block from the power plant by the same name. The apartments in Tony's building have a clear view of the Ravenswood Generating Station, which dominates the skyline alongside the Queensboro Bridge.

Sandwiched between these two gigantic power production facilities, Astoria had become known as an "asthma alley" for its poor air quality and incidence of the number of days children missed school because of asthma. It joined its neighbors, Harlem and the South Bronx, in this dubious distinction. Tony describes the meeting that propelled them to start a community movement for clean air:

> *I'm a lifelong resident of Astoria, and I've been active in the community since I've been here. This whole thing started when we went to a community meeting because we heard they were going to put in another power plant. And at that meeting the gentleman from the company got up and said, "well, there's no way for you people to stop us, we got it from the siting board; we're going to put this power plant in no matter what you do." And at that point, I turned around and said, "Well, that's news to us."*[20]

Tony and many of his neighbors launched a community-wide campaign that began right after this meeting—the co-op boards, the homeowner's associations, the environmental groups, state legislators—ultimately building a coalition that included many state and city officials and a national legal advocacy group.

He describes how their campaign shut down the Poletti Power Plant and stopped the proposed new one from being built:

> *I talked to our board members, and I said I'd like to organize all the civic associations and co-op boards and condo boards and homeowners associations in the area . . . to organize and see what we could do about this. We're taking the brunt of it right now, with all the power plants in our area.*[21]

I asked him to elaborate on how they got people interested. After all, fighting a new power plant does not seem like an easy thing to win:

> *I was telling them . . . the 11105 area code had the highest throat cancer rate in our area. We had the highest asthma rate. That got people involved. The mothers of all the children in the schools, they heard that . . . they got involved. And I think that's what you got to do . . . people just think the electricity just flows and, you know . . . where is it made? . . . Meanwhile, you breathe in all that pollution. You're living nice, but you're going to die sooner. Pollution doesn't discriminate.*[22]

Astoria Generating complex of power plants on one side, and on the other side the Ravenswood Power Plant, located on the East River right underneath the Queensboro Bridge: *"It's not a case of 'not in my backyard' . . .*

even though we do provide more than half of the power for New York City, right here, in this area with these five power plants. We created a coalition organizing a cleaner environment. It took us . . . eight, ten years. But we did it."[23]

It took constant protests and activism at local and state levels, educational events throughout the city, media stories such as one featured in New York Magazine and other mainstream press, and a lawsuit by the National Resources Defense Council—but they won. They stopped the new power plant from being built, won an order decommissioning the Poletti Power Plant, and won a promise from the utility to clean up the remaining power plants.[24]

For an even more detailed account of shutting down the Poletti Power Plant, see the 2018 article by Rebecca Bratspies, in which she carefully re-captures and recreates how community activists shut down New York City's most polluting power plant. In her words: *"The story of shutting down the Poletti is a tale worth telling, and a potential template for successfully instantiating the human right to a healthy environment."* Addressing the environmental justice impact and legal aspects, Bratspies *"gleans lessons from the victorious legal and regulatory campaign and suggests how those lessons can be usefully deployed by environmental and human rights advocates going forward."*[25] Or, she published a 2020 article detailing the sustainable development lessons from this campaign, concluding:

> *The message is clear: things will change only if ordinary people demand it. Past advocacy efforts can offer guidance for what a successful claiming of sustainable development might entail. There is much to learn from close examination of successful, socially-oriented environmental advocacy—from what successful advocates/activists did on the ground, and how they shaped their environmental claims.*

The full article examines lessons from the Astoria campaign that offer a template for contemporary environmental and sustainable development initiatives, particularly those linking environmental concerns with local economic development.[26]

Astoria Shares Its Win—New City Councilmember Leads Environmental Committee

Right after the Astoria residents group won this legal battle, they elected a City Council member who pledged to keep the fight going and make sure the utilities live up to their promises. Costa Constantinides took office in 2014 and became chair of New York City Council's Environmental Committee.

To say Costa has continued the effort to clean up the air in Astoria would be a massive understatement. He has led the passage of 2019 legislation regarded as the strongest in the world to control building emissions—Local Law 97, otherwise known as "New York City's Green New Deal." Many have called this the strongest, most ambitious law for improving the efficiency of buildings in the nation and possibly the entire world.

Two years before that would happen, Costa spoke to me in an interview right in front of the Astoria Energy compound, in a small park running alongside the East River. He shared how Poletti had been affecting the community:

What you don't see [in this compound] is a plant that used to be there. What used to be here was the Poletti Power Plant, which was the dirtiest plant in New York state, and which was decommissioned in 2010 or so through the advocacy of this neighborhood. It was the worst technology. It was blowing right in these residents' faces for decades. We got that decommissioned, and then finally, in about 2015, fully taken away. It was dismantled because . . . we always thought if there was an opportunity to reopen it, they were going to take it. So we fought very hard with the Power Authority to make sure that it came down.[27]

When he was elected, Costa made it one of his priorities to push the utilities to stick to their promises. He also initiated many initiatives for clean air, including passing several other green buildings laws that created stringent restrictions for reducing emissions from buildings. For example, completely banning the use of Number 6 oil, the toxic sludge used for heating and which has been mentioned at the beginning of this chapter. Despite previous City Council legislation governing building emissions and limiting the use of Number 6 fuel oil, the volume of large buildings in New York City makes it a constant battle to ensure that this legislation is enforced. Additionally, building owners who do not want to spend the money to convert to newer, cleaner systems were often able to find compliance loopholes. Some simply paid fines and kept polluting. The 2017 law restricted the emissions from these buildings even more.

Costa explained how this worked:

At the end of the day, even some of the larger building owners recognize that climate change is real. It has to be dealt with in a significant way. And the consequences of not dealing with it . . . seeing emissions and climate models of places that we know where our families live, where we live—like the Rockaways, Coney Island. Lower Manhattan—disappearing off the map. A hundred thousand New Yorkers potentially being climate refugees in their own city. These are mostly environmental justice communities. Even those who you'd think are ideologically opposed to this bill are somewhat tempered in their opposition.

We had all stakeholders involved. We're expecting that we're going to be able to get to a really good place.[28]

Focusing again on Number 6 oil, the culprit responsible for a massive proportion of the city's pollution, he elaborated on how the legislation would work:

Number 6 oil is something that's already banned, but this would actually look at reducing depending on the starting point. So the worst emitting buildings, like Trump International Tower which has a 12% energy star rating.[29] *That building would, by 2022-23, have to make a significant cut in their emissions. Then by 2030, they'd have to come down another significant number. They're going to have to change the way they do business. It's about looking at your building systems. Changing not only from Number 6 oil, but putting more solar, wind power, geothermal, and looking at alternative energy sources for your buildings. This is going to be a paradigm shift for building management in the City of New York. And it's way overdue.*[30]

We discussed the complexity of trying to reduce air pollution, his motivations, and how he sees these laws having a broader impact:

I grew up in this neighborhood. We create over 55% of the city's power here. We have the Grand Central Parkway that runs down the middle of it. We have the LaGuardia airport, which is another source of pollution. We have a lot of large buildings. We have a sewage treatment plant, we have a combined sewage overflow that flows into our waterways. We have an "embarrassment of riches"—and those are real costs to our neighborhoods.

Some politicians take up green initiatives around election time, with well-versed talking points on sustainability. Public skepticism about their commitment can be high. Talking to Costa felt more like talking to your neighbor, who cared just as much as you do, about the soot on your windowsill (and in your lungs) and the smell from the wastewater treatment plant when the wind blew the wrong way. I asked where this came from for him:

We see asthma rates that are through the roof here, higher than the borough average for E.R. visits, higher than the borough average for hospitalizations. It's personal to me . . . Growing up in this neighborhood and, and seeing a community that has always fought, growing up with this legacy of residents who were saying they are not accepting the status quo . . . who fought to get rid of the Poletti power plant, who expect better and want better for the next generation. Now that I'm raising my family here, I'm doing the same. So we're all in this together.[31]

We turned the interview toward the power plant, which has become a ubiquitous part of the skyline. I asked about the many protests and rallies and speeches he has made in this very spot:

It's on days like this where you see the stacks and [it looks like] *there's nothing coming out of them.* [But] *there's still emissions coming out of there every day. So early in this year, when we had a rally, we made sure it was a cold day, so it would crystallize. You'd see the white smoke. So then people would realize, Oh, this is the pollution that populates our neighborhood that makes our kids sick. That increases our asthma rates.*[32]

When asked to elaborate on New York City's role, he explained what is special about this transition happening in America's largest city:

We can set market trends that will push back on climate change . . . when you look at biofuel, the American Petroleum Institute spent close to $200,000 to kill our bill here in New York City. Because they knew it wasn't about the millions of gallons that we weren't going to burn in petroleum here. It was also the fact that the largest city in the Northeast was going to no longer be burning millions of gallons of petroleum . . . what would that mean for other cities and other sources around New York City?

We're going to create a market driver that's going to drive other cities to say, well, you know, if our largest customer in New York City is green, then why . . . are we going to make number six oil now? Why are we going to move forward on fossil fuel? So we're going to set a trend.[33]

In addition to banning and limiting the dirtiest fuels, he has included the installation of solar panels on city-owned buildings in his district's budget: *Just this past year, I funded the first two public buildings in our neighborhood to be solar panels. So PS 122, which is a few blocks away from here, will have solar panels and will be able to teach kids about how solar panels work.*

By 2019, Costa had funded six schools to have solar panels, along with the Steinway library and other city buildings in his district. His legislative track record includes bills authorizing the installation of small wind turbines in New York, a proposed bill for large wind turbines which was approved by the City Council, and introduced legislation to support the installation of utility-scale battery storage systems on city buildings. His record includes dozens of other sustainability-related laws, including passing another 2019 bill mandating the City Council to create a plan to close all of the fossil fuel power plants within New York City, including the natural gas-powered ones.

This is all in addition to leading the effort to pass Local Law 97, which was part of the Climate Mobilization Act, the official name of the New York City's Green New Deal—regarded as the largest carbon emissions reduction mandated by any city in the world. The Climate Mobilization Act set strict climate-focused laws. And, it was necessary: many similar goals had been established on paper at city and state levels. However, the progress toward them was sluggish. A 2019 report by the state found that despite many climate-focused initiatives limiting emissions, they only succeeded in reducing emissions by 8 percent in the 25 years from 1990 to 2015.[34]

As the lead voice to pass 2019's Climate Mobilization Act, Costa took the spirit that got the Poletti Power Plant shut down and dismantled, and amplified it city-wide. New York City's Green New Deal was a coalition effort, backed by environment groups, labor organizations, community-based groups, and dozens of other stakeholders, who made their desire to see serious progress toward climate action heard throughout the city, the state, and by the entire world.[35]

In looking at this kind of track record, which started on a community level and developed into legislation governing the entire city, we also need to look at the actual, day-to-day impact of a City Council bill. Local Law 97 is strong and ambitious, but it also expands on restrictions that had existed in previous laws.

The question remains: If New York City had previous laws, and they didn't go far enough, why would this one be different? What could be the real impact of this City Council bill on regulating pollution? I also took this question to the founder of Columbia's Sabin Center for Climate Change. He responded:

I think the most important climate-related bill passed by the City Council in recent years is Local Law 97, which imposes aggressive requirements on buildings in NYC to reduce their greenhouse gas emissions. Local Law 97 is a substantive law that will have a significant impact. Resolutions that just call on agencies to consider or prioritize specific issues have less of an impact; they do not drive change nearly as directly.

He also pointed me in the direction of a resource their Center has made available to the public—The New York City Climate Law Tracker[36]—a tool to track these many laws, their milestone dates, action items, responsible entities, the status, and even find related articles. Looking at it, on the docket at the time of writing are several items and notes for Rikers Island; with due dates in summer 2021, including a date an advisory committee is supposed to meet (August). This kind of detailed, public-facing resource is of great value, and I include it here in hopes that it will be used extensively by other researchers, by students, and by those who will track, observe, and ensure these laws are put into practice.

For an elected official, it takes incredible patience, time, and strategy to pass laws that result in impact like the owner of a skyscraper finally spending the money to change their dirty fuel oil heating system to a cleaner one. For Costa, as indicated in the quote above, this drive comes from a personal place. He saw his community win a fight to shut down a dirty power plant, and he also knows firsthand the impact of polluted air on families; his son has asthma and takes several medications daily to control it. When Costa fights for clean air, he stands alongside other families and community members affected by asthma and respiratory illnesses, whose children, like his own, will bear the brunt of pollution for their entire lives.

ASTORIA POWER PLANT EXPLOSION—CAN WE JUST SHUT THEM ALL DOWN?

A community initiative in the northwest corner of Queens inspires a local resident to run for City Council, and ends up impacting all 20 million people in the New York City metro area through the legislation he introduces and the coalitions he builds to pass them. Does it take deeply personal motivation like this for an elected official to stick with the challenges of continually passing more substantial bills, until they finally enact a law that has substantive impact?

It seems the answer is yes. A few months prior to passing the Green New Deal bill, Costa had also introduced another sweeping bill in response to the "#AstoriaBorealis." In December 2018, a massive explosion lit up New York City skies with a brilliant turquoise glow. It lasted for about 20 to 30 minutes and was bright enough to see from New Jersey. People on social media speculated that aliens were invading. If you were close to the plant, you could hear the eerie electric whine. No one knew what it was.

A transformer had blown at the Astoria Energy complex, with its five power plants, located across a two-lane city street from family homes, and right between Rikers Island and LaGuardia airport. It caused blackouts at both LaGuardia and the Rikers Correctional Facility.[37,38] Families were home for the holidays, and their houses shook. This neighborhood is densely packed with homes, and there is a gas station across the street from the power plant. For people who lived right there, it wasn't beautiful or exciting or otherworldly—it was terrifying.

The turquoise glow came from an electric arc caused by the transformer exploding. An arc is caused by electricity flowing through the air rather than through a wire or its intended source. They are caused by faulty equipment—in a home they are caused by frayed wires and corrosion. They are extremely dangerous because of the intense concentration of energy in the arc. They can jump to a different conductor, changing paths in mid-air. The heat can cause fires.[39]

Luckily, no one was hurt this time. The utility claimed their equipment malfunctioned. Video footage from social media showed that Queens residents could walk right up to the standard-issue chain link fence separating the power plant from the street and sidewalk—not much protection from malfunctions and explosions.

In the first week of January 2019, Costa held a press conference in front of the Astoria Energy complex to address how the turquoise explosion highlights the need to reduce and eliminate fossil fuel-burning power plants. At the press conference, he introduced a bill to propose the City Council start looking at how to get rid of all 24 power plants within city limits. In the coming months, City Council members, including Costa, questioned Con Edison executives about the explosion and how to prevent it, although those hearings ultimately did not provide any answers, nor did Con Edison accept that there is a danger to community members from the power plants.[40,41]

While shutting down all of New York City's 24 fossil fuel-burning power plants may seem extraordinarily ambitious both politically and technically, shutting down even one coal-burning power plant also seems ambitious and impossible—yet, it happened. When the first power plant in the U.S. was built by Thomas Edison in 1882—a coal-burning central power station called the Pearl Street Station[42]—it was unheard of to think the entire world would one day have 24/7 electrical power in every corner. Yet, that also happened. This chapter's focus on *how* America's largest city will realize its promises and commitments for clean energy looks at technical solutions. It also looks at human solutions, as shown in the personal commitment of people like Costa, Rebecca, and Tony, who are all patient yet forceful, and diligently persevere to make the impossible become the actual.

COMMUNITY SOLAR

There are promising new developments for creating new distributed generation sources. I expanded the look at New York City's transition to renewables beyond City Council legislation, and also looked at what was possible from a consumer perspective. Very few New Yorkers are like that Staten Island family from chapter 2 of this book; most New Yorkers do not own their own roof. If you rent or live in an apartment, how can you go solar? Can only homeowners be part of the clean energy economy? What if you are a homeowner, but for any of a variety of reasons, your roof is not suitable for solar—for example, when there is not enough sun, or the roof's structure cannot fit the panels?

Community solar programs have emerged in New York and expanded throughout the region. They create one way for the people who do not have access to rooftop solar so that they can be a part of the clean energy revolution. As discussed elsewhere in this book, there are many versions of community solar. Each chapter discussed a different way community solar works in varied locations.

Here is how it works: a solar energy company installs and maintains a solar farm on large rooftops throughout the city. In New York City, these sites are primarily warehouses, garages, factories, houses of worship, and other large flat rooftops. Often, they are located in industrial areas with little to no tree coverage. These are ideal locations for solar farms. The building owners receive revenue by leasing their roofs to the solar company. Consumers buy in by signing up for a contract, the company switches them, and they pay through their regular electric bill. The power created by the solar farms goes into the standard electric mix through an agreement with the utility. The consumer's bill stays exactly the same. However, their source of power has switched to community solar. (The program I signed up for guaranteed electricity rates at 5 percent or more below whatever the regular consumer rate was at the time.)

This is an emerging and developing model for power production, and it has its share of complications. First and foremost, it was confusing to consumers to understand how it worked. It is complicated on a regulatory level and requires collaboration between the solar company, the utility, and the regulatory bodies. As with any type of new business models, finding where community solar fits in the ecosystem of ways to access renewable energy will take time.

At the moment, community solar is a tiny drop in the massive bucket of New York City's electrical consumption. However, as with many of the smaller projects described throughout this book, community solar in New York does far more than provide electricity to a few hundred homes. It shows the unrealized potential for urban solar, it creates a working business model to expand access to solar beyond homeowners, and it forges a new way for businesses to use solar for economic development.

A critical mass does seem to be growing. The community solar program I switched to had fewer than five sites when I first met them in 2018. I checked on their progress about a year after signing up, and they now have about 10 sites on their website. At the time of writing this, 2 years later, they have 54 sites in several states. Clearly, the public welcomes this new way to buy energy.

New York, NY

RENEWABLE RIKERS

As mentioned in other parts of this chapter, in October 2019, New York City leaders finally, after years of activism, voted to close down the infamous prison on Rikers Island. This initiative was a long time in coming and originated from the Independent Commission on New York City Criminal Justice and Incarceration Reform, a body formed to study the closure of the island and to suggest possibilities to repurpose it. One of the reasons cited for the closure of the island is its inaccessibility; the only way to reach it is by land, from a single bridge in Astoria. There is one bus that goes to Rikers Island, and a cab would take almost an hour even from many parts of Queens, much less another borough.

This inaccessibility was challenging for everyone. Families could not visit their loved ones without taking an entire day to travel, and legal representation found it equally time-consuming. Even the correctional officers who worked there had trouble getting there.[43]

More so, closing Rikers is also restorative justice. In a sweeping 148 page report from the Independent Commission, referred to as the Lippman Report, stated: "We believe that a twenty-first century justice system must acknowledge the multiple harms that incarceration, and Rikers Island in particular, has caused hundreds of thousands of New Yorkers, their families, and their communities. And, it must acknowledge that these harms fall disproportionately on communities of color. To heal and restore hope, a jail must become a last resort rather than the path of least resistance."

This influential Lippman Report did not stop at its recommendation to close Rikers Island correctional facility. Given the premium on any kind of land in New York City, they also considered the use of Rikers Island and what it could become, particularly in light of other city priorities focused on equity and sustainability. One headline summed up the potential for a land grab: "Available: A 415-Acre Island with Manhattan Views. What to Build?"

The Lippman Report outlined several possible ways to repurpose Rikers Island, which were in line with sustainability and resiliency:

> Rikers Island presents an opportunity to address both challenges by providing open land area for a large-scale solar energy installation and a strategic site for an energy storage system. While both solar arrays and battery storage are modular technologies that could exist at a range of sizes on Rikers Island, in consideration of other potential uses, the estimated high-end capacity that could be sited is approximately 90 megawatts of solar production—enough to power nearly 25,000 households—and 300 megawatts of energy storage. Growing the City's solar capacity would reduce its reliance on fossil fuelproducing power plants. The ability to efficiently store power generated by renewable sources

would also help eliminate the need to build and run expensive conventional power plants to meet peak demand.[44]

This proposal to convert Rikers Island into a solar farm met with strong public support. It also included the need for job training focused on communities that have been the most impacted by Rikers Island and criminal justice inequities.

Costa introduced "Renewable Rikers" to the New York City Council in January 2019, a trio of bills that would move forward this plan. There are certain specifics that make it a highly strategic area of development. As mentioned above, the solar power and battery storage would be sufficient to meet peak demand. Currently, peak demand is often met by turning on what are called "peaker plants," which are smaller generation stations within primary power plants. Peakers are turned on when there is high demand. Peakers in New York City often burn oil and are much dirtier than the natural gas-powered main generation stations. By replacing this one piece of the electrical system, the Renewable Rikers plan would make a significant dent in one of those loopholes that allows for burning the dirtiest and most polluting fuels.

In an interview where he first spoke to me about this, Costa described the plan:

> As we look at Rikers Island Prison closing, we strongly feel that we can replace Rikers Island with renewable energy. We can have a solar farm there that can replace the energy of all of the peakers in New York City. We can build a wastewater treatment plant to replace the wastewater treatment plants that are mostly cited in environmental justice communities throughout the City of New York. That opportunity—by closing Rikers—will not only create a social justice revolution in the City of New York, it would create an environmental justice revolution in the City of New York, by allowing us to take that land and turn it into a real benefit to a cleaner grid and a betterment of neighborhoods.

This is not an environmental fantasy—it is a viable technical reality coming into use in other cities. Utilities are starting to use batteries to replace peaker plants, and, as they start to gather data on performance and cost, they are acknowledging that batteries not only have the capacity to replace peakers entirely, there are many advantages to using utility-scale batteries over peakers. These include more flexibility in sizing, technical flexibility, greater efficiency, and cost,[45] and the energy industry media are starting to address how batteries will replace peakers.[46]

Reclaiming Rikers Island following this plan would support New York City in reaching its ambitious emissions reductions, improve resiliency, and be a pragmatic step toward modernizing critical city infrastructure. It would also be a symbolic step to address Rikers racist history. In a Loyola Law Review article, Rebecca Bratspies projects how Renewable Rikers will create tangible benefits for environmental justice communities:

> By converting part of the island into a wind/solar generating and storage facility and part into a state-of-the-art modern wastewater treatment facility, the City could reduce reliance on (and possibly shutter) the old, dirty, peak-load generating facilities that are located in environmental justice communities. Such a step would dramatically improve air quality in those communities and create green jobs; potentially rerouting the school-to-prison pipeline into a school-to-green jobs pipeline. Moreover, the prime waterfront space currently occupied by these noxious uses could be converted into accessible, much-needed green spaces in currently underserved communities—an additional restorative step that would also promote environmental justice.[47]

With all of the backing and research showing that this would be a very realistic plan, Costa and many others took it to the public. At a public meeting held in Jackson Heights, Queens in April 2019, Costa and dozens of other elected officials and community leaders ranging from state senators to leaders of New York City environmental and community development nonprofits, to the U.S. representative Alexandria Ocasio-Cortez (whose district includes Queens and the Bronx, and encompasses Rikers Island) gathered for a public town hall on Renewable Rikers.

The room was packed from wall-to-wall. Outside, protesters waved signs on the sidewalks, shouting "America first" and in favor of keeping Rikers open. Inside, the crowd was energetic. Rebecca moderated the event. As with other communities described in this book, when people see how energy affects their immediate lives, they understand and become very engaged. Panelists included formerly incarcerated individuals who could be trained to reenter the workforce as solar installers sketched out how Renewable Rikers would affect them.

This is far more than a progressive dream. In January 2020, Renewable Rikers was on their docket for the New York City Council Spring 2020. When the Covid-19 pandemic struck, and the City Council had to focus on surviving the pandemic, Renewable Rikers was rescheduled, eventually hitting the City Council docket in February 2021. It was passed by the council and signed by the mayor. There were also strict dates set for implementation of steps like turning over land from the correctional facility to city administration, and creating an advisory board to monitor progress and compliance, with those milestone deadlines starting in July 2021. It also set a 1-year deadline to create an energy plan for using renewable energy with battery storage.

In the press release, Costa noted:

> The 413 acres of Rikers Island have, for far too long, embodied an unjust and racist criminal justice system. Far too many New Yorkers found themselves caught in a cycle of over-policing and over-incarceration symbolized by an island named for the family of a slave catcher. Now, however, we will have a golden opportunity to put the principles of the Green New Deal into practice with the Renewable Rikers Act. These bills will offer the city a pathway to building a hub for sustainability and resiliency that can serve as a model to cities around the world.

CONCLUSION

In fall 2021, New York City will elect a new mayor and will have sweeping City Council change, with over 50 percent of the current members coming up to their term limits. Costa is one of those, and as he moves on to become CEO of a youth organization, a new Astoria City Council member will step in, and a new Environmental Committee chair will be chosen. Tony is still on the board of his co-op, and Rebecca continues to write and publish about urban environmental reform, and a tenacious city-wide coalition to support these initiatives now exists.

But at the moment, those 24 power plants are still there, chugging pollution into New York City's air, causing asthma, cancer, and respiratory illnesses. Along with them are a fast-growing number of community solar projects, a handful of wind turbines, and powerful legislation creating a pathway to change. People from criminal justice reform, environmental justice fields, and renewable energy fields are united in their commitment to realize New York's Green New Deal and Renewable Rikers. The economic impact of the coronavirus pandemic and lockdown is still unfolding, with the New York City and state suffering staggering financial losses. People are also living with the long-term effects of COVID 19. Clean air and green jobs are more necessary than ever.

At one point during an interview, Costa had been telling me about several successful pieces of legislation, which spoke to a combination of visionary persistence and stubborn resolution, but also sounded exhausting, and I asked how he maintained this energy. His answer speaks volumes about the strategy and persistence for making these bills into actuality:

> You know, when you climb a mountain, there's another mountain. Then another mountain. We know we need to start looking for ways to move away from fossil fuel electric generation. How can we get power plants out of the

city? How do we incentivize battery storage and renewable energy? I'm not going to stop.[48]

As New York City and the state pick up the pieces after Covid-19, we are seeing our next mountain. With a starting point like this, this movement is never going to stop.

NOTES

1. U.S. Census Bureau (2019). *New York City*, New York. Retrieved from https://censusreporter.org/profiles/16000US3651000-new-york-ny/
2. U.S. Census Bureau (2019). *New York-Newark-Jersey City*, NY-NJ-PA Metro Area. Retrieved from https://censusreporter.org/profiles/31000US35620-new-york-newark-jersey-city-ny-nj-pa-metro-area/
3. U.S. Census Bureau (2019). *New York City*, New York.
4. New York ISO. "Gold Book - 2019 Load and Capacity Data." *New York Independent System Operator*. 2019. https://www.nyiso.com/documents/20142/2226333/2019-Gold-Book-Final-Public.pdf/a3e8d99f-7164-2b24-e81d-b2c245f67904
5. New York City Energy Policy Task Force. "New York City Energy Policy: An Electricity Resource Roadmap." *New York City Energy Policy Task Force*, January 2004. http://www.nyc.gov/html/om/pdf/energy_task_force.pdf
6. Bratspies, Rebecca M. "Renewable Rikers: A Plan for Restorative Environmental Justice." Available at SSRN, July 24, 2020.
7. Gerrard, Michael. Email to author. August 5, 2020
8. Gerrard, Michael. Email to author. August 5, 2020
9. New York State Independent System Operator. "What We Do." *NYISO.com*, accessed December 29, 2020. https://www.nyiso.com/what-we-do
10. New York State Reliability Council. "About New York State Reliability Council." *Nysrc.org*, accessed December 29, 2020. http://www.nysrc.org/
11. New York ISO. "Power Trends 2021: New York's Clean Energy Grid of the Future." *New York Independent System Operator Annual Grid and Markets Report*. 2021. https://www.nyiso.com/documents/20142/2223020/2021-Power-Trends-Report.pdf/471a65f8-4f3a-59f9-4f8c-3d9f2754d7de
12. U.S. Energy Information Administration. "Today in Energy: New York's Indian Point Nuclear Power Plant Closes After 59 Years of Operation." *Eia.gov*. April 30, 2021. https://www.eia.gov/todayinenergy/detail.php?id=47776
13. U.S. Energy Information Administration. "Indian Point, Closest Nuclear Plant to New York City, Set to Retire by 2021." *Eia.gov*, February 1, 2017. https://www.eia.gov/todayinenergy/detail.php?id=29772
14. U.S. Energy Information Administration. "Today In Energy: New York's Indian Point Nuclear Power Plant Closes After 59 Years of Operation." *Eia.gov*. April 30, 2021.

15. Roston, Eric and Wade, Will. "As Indian Point Goes Dark, New York Races to Swap Nuclear With Wind." *Bloomberg Green, Energy & Science*. April 30, 2021. https://www.bloomberg.com/news/articles/2021-04-30/indian-point-nuclear-plant-shuts-down-and-new-york-races-for-wind-power

16. Haag, Matthew. "N.Y.C. Votes to Close Rikers. Now Comes the Hard Part." *New York Times*. October 17, 2019. https://www.nytimes.com/2019/10/17/nyregion/rikers-island-closing-vote.html

17. Bratspies, Rebecca M. "Renewable Rikers: A Plan for Restorative Environmental Justice." Available at SSRN, July 24, 2020. https://ssrn.com/abstract=3660113 or http://dx.doi.org/10.2139/ssrn.3660113

18. Hendrick, Daniel. "Queens' Bad Air Worsens—Sharp Rise In Toxic Emissions From Power Plants." *Queen's Chronicle*, July 10, 2003. https://www.qchron.com/editions/queenswide/queens-bad-air-worsens-sharp-rise-in-toxic-emissions-from/article_c833c111-fa77-5c32-8086-f4b8c2d4e18a.html.

19. Pérez-Peña, Richard.. "State to Close Queens Plant That Is Biggest Polluter in City (Published 2002)." *New York Times*, September 5, 2002. https://www.nytimes.com/2002/09/05/nyregion/state-to-close-queens-plant-that-is-biggest-polluter-in-city.html

20. Gigantiello, Anthony. "How to Power a City." Interview by Melanie La Rosa. August, 2017.

21. Gigantiello, Anthony. "How to Power a City." Interview by Melanie La Rosa. August, 2017.

22. Gigantiello, Anthony. "How to Power a City." Interview by Melanie La Rosa. August, 2017.

23. Gigantiello, Anthony. "How to Power a City." Interview by Melanie La Rosa. August, 2017.

24. Pérez-Peña, Richard.. "State to Close Queens Plant That Is Biggest Polluter in City (Published 2002)". *New York Times*, September 5, 2002. https://www.nytimes.com/2002/09/05/nyregion/state-to-close-queens-plant-that-is-biggest-polluter-in-city.html.

25. Bratspies, Rebecca M. "Shutting Down Poletti: Human Rights Lessons from Environmental Victories." Wisconsin International Law Review. December 28, 2018. Available at SSRN: https://ssrn.com/abstract=3307512 or http://dx.doi.org/10.2139/ssrn.3307512

26. Bratspies, Rebecca M. "Public Housing, Private Owners: Sustainable Development Lessons from the Fight to Shut the Poletti Power Plant." Cambridge Handbook of Environmental Justice and Sustainable Development, May 6, 2020. Available at SSRN: https://ssrn.com/abstract=3593946 or http://dx.doi.org/10.2139/ssrn.3593946

27. Constantinides, Costa. "How to Power a City." Interview by Melanie La Rosa. June, 2017.

28. Constantinides, Costa. "How to Power a City." Interview by Melanie La Rosa. June, 2017.

29. New York City. "Benchmarking and Energy Efficiency Grading." *NYC.gov*, accessed December 29, 2020. https://www1.nyc.gov/site/buildings/business/benchmarking.page

30. Constantinides, Costa. "How to Power a City." Interview by Melanie La Rosa. June, 2017.
31. Constantinides, Costa. "How to Power a City." Interview by Melanie La Rosa. June, 2017.
32. Constantinides, Costa. "How to Power a City." Interview by Melanie La Rosa. June, 2017.
33. Constantinides, Costa. "How to Power a City." Interview by Melanie La Rosa. June, 2017.
34. McKinley, Jesse, and Brad Plumer. "New York to Approve One of the World's Most Ambitious Climate Plans." New York Times, June 18, 2019. https://www.nytimes.com/2019/06/18/nyregion/greenhouse-gases-ny.html
35. New York City League of Conservation Voters. "New York City Climate Action Tracker." *Climatetracker.nylcvef.org*, accessed December 29, 2020. https://climatetracker.nylcvef.org/
36. Sabin Center for Climate Change Law. "New York City Climate Law Tracker." *Climate.Law.Columbia.Edu*. Sabin Center for Climate Change Law, Columbia Climate School. Accessed June 15, 2021. https://climate.law.columbia.edu/content/nyc-climate-law-tracker.
37. Stevens, Matt, Rick Rojas, and Jacey Fortin. "New York Sky Turns Bright Blue After Transformer Explosion." *New York Times*, December 28, 2019 https://www.nytimes.com/2018/12/27/nyregion/blue-sky-queens-explosion.html
38. Kaufman, Alexander. 2019. "Power Plant Accident Casts New Light On New York'S Dirty Fuel Addiction". *Huffpost.Com*. https://www.huffpost.com/entry/transformer-explosion-nyc-fuel-addiction_n_5c25c357e4b0407e9081305a?guccounter=1.
39. Atlanta Electrician. "White Paper: What Is Electrical Arcing and Why Is It Dangerous?" *Atlanta ElectricianAtlanta.com*. December 28, 2017. https://www.electricianatlanta.net/what-is-electrical-arcing-and-why-is-it-dangerous/
40. Metro New York. "No Easy Answers on Astoria Borealis." *Metro.us*, February 11, 2019. https://www.metro.us/no-easy-answers-in-astoria-borealis-transformer-fire-hearing/
41. Shahrigian, Shant. "ConEd Grid Guys Grilled by City Council For 'Inadequate and Laughable' Response to Summer Blackout." *New York Daily News*. September 4, 2019. https://www.nydailynews.com/news/politics/ny-coned-city-council-corey-johnson-grilling-20190904-v7crvydvc5ebdpe43rcj7xki4m-story.html
42. Harvey, Abby, Aaron Larson, and Sonal Patel. "History of Power: The Evolution of the Electric Generation Industry." *Power Magazin/powermag.com*, accessed December 29, 2020. Originally published October 2017, updated December 22, 2020. https://www.powermag.com/history-of-power-the-evolution-of-the-electric-generation-industry/
43. Independent Commission on New York City Criminal Justice and Incarceration Reform. "A More Just New York City." *Independent Commission on New York City Criminal Justice and Incarceration Reform*, April 2017. https://www.morejustnyc.org/reports

44. Independent Commission on New York City Criminal Justice and Incarceration Reform. "A More Just New York City." Page 111.

45. Greentech Media. "How Storage Can Help Get Rid of Peaker Plants?" *Greentech Media*, June 28, 2010. https://www.greentechmedia.com/articles/read/energy-storage-vs-peakers

46. Newbery, Charles. "Energy Storage Poses a Growing Threat to Peaker Plants." General Electric, *Transform blog*, October 1, 2018. https://www.ge.com/power/transform/article.transform.articles.2018.oct.storage-threat-to-peaker-plants

47. Bratspies, Rebecca M. "Renewable Rikers: A Plan for Restorative Environmental Justice." Available at SSRN, July 24, 2020. https://ssrn.com/abstract=3660113 or http://dx.doi.org/10.2139/ssrn.3660113

48. Constantinides, Costa. "How to Power a City." Interview by Melanie La Rosa. December, 2018.

Chapter 5

Colchester, VT
Green Mountain Power

INTRODUCTION

In Colchester, Vermont, just a few miles northeast of Burlington, America's first benefit corporation utility has set the gold standard for innovation in the production and distribution of electricity. Green Mountain Power earned the title of "America's most innovative utility" by ushering in important technical innovations like distributed generation, wind farms that provide revenue to nearby towns, and being a frontrunner in using Tesla Powerwall batteries. Throughout this book, I discuss how communities lead the transformation and evolution of energy systems. Green Mountain Power (GMP) is also leading a different and related type of transformation: how a business makes, distributes, and talks about power.

While asking about Green Mountain Power's technical progress, I received answers about far more than technology—they were equally detailed and descriptive about people as they were about machinery. They addressed corporate culture and the relationship between customers and the utility. They looked at utility innovations inside the boardroom as well as outside on the grid and at power plants.

Utilities and power companies are big, wealthy businesses, typically with a staid image about technology, faceless infrastructure, and indifference to anything other than the bottom line. Not this one—Green Mountain Power's focus on and acknowledgment of the importance of customer service is perhaps their most important innovation, as it opens up the door to future developments. In Vermont, utilities function as regulated monopolies; customers must purchase electricity from the one utility in whose service area they reside. In comparison, the rest of New England has, in recent years, changed the electric industry's business structure to allow more customer choices.[1] GMP focused on customer service even when their customers had no other option but to purchase power from them.

If you are interested in clean, renewable energy, it is reasonable to expect you will have read many accounts of how utilities work with communities. As you read through this book, you will find many different examples of these interactions. In some, utilities acted in ways to be destructive to the environment and community and indifferent to customers' needs, such as in Puerto Rico, Highland Park, and New York City (at one point). In other examples, utilities collaborated with other parties to bring clean energy projects to life, such as in Las Vegas, Atlantic City, and New York City (at another point).

In Vermont, Green Mountain Power did far more than collaborate—they led this transformation. They also did something less visible and equally impactful: they changed their behind-the-scenes culture and business structure to support ongoing innovation.

CHAPTER SUMMARY

A theme that runs throughout this entire book is the idea of community-informed responses to energy issues. In many other chapters, "community" means a neighborhood or perhaps a type of identity. There are also business communities for energy. Other chapters throughout this book look at projects led by individuals, some very small in scale. However, to fully sketch a picture of energy transformation, including a utility's perspective is crucial. And what better utility to learn from than the one nationally recognized as the absolute forerunner in the field?

This chapter explores aspects of that perspective through an interview with Mary Powell, former CEO of Green Mountain Power (GMP). During her tenure, GMP became a B-Corp and achieved a customer satisfaction rating above 90 percent. As states and cities set goals to use high proportions of renewable energy in their electrical supply, GMP has already done this. They currently get 64 percent of their energy mix from renewable sources. They have used over 60 percent renewable energy in this mix for many years, with 95 percent of it carbon-free. Their goal is to use 100 percent renewable energy by 2030. This brief chapter provides insight into how they did this, including the shift in the business model and conceptual approach. This chapter is not a comprehensive look at GMP's many accomplishments, which are too numerous for these few pages. It is, instead, a sketch of how focusing on *people* translated into transforming energy systems.

METHODOLOGY

During my background research, I asked Karl Rábago at the Pace Energy and Climate Center whether he knew a utility that he thought would be willing to speak with me for the film. He immediately suggested Mary Powell, CEO of Green Mountain Power.

At the point in the research when I asked for this referral, I had already spoken with a fair number of utilities as well as with companies that worked closely with utilities. (See chapter 10 on Las Vegas for a detailed account of a city working with a utility.) Some utilities and government officials I spoke with chose not to respond. Others spoke to me briefly but off the record, saying they could not respond officially. Others responded with answers that were very deliberate and so carefully worded that they were entirely opaque. Including those responses would have done little to facilitate public understanding of clean energy. Green Mountain Power, along with NV Energy, responded with interest and receptivity. Neither asked to review any of my materials before being released. Both were responsive to my questions, provided me with additional information, and seemed to value this research as an opportunity to share their work and perspective with the public. This open sharing of information was refreshing.

The interview with Mary Powell was conducted in October 2018. Mary helmed Green Mountain Power for 12 years, from 2008 to 2020. This interview was also informed from a follow-up interview with a Green Mountain Power engineer, and I was provided access to film at a solar farm on a landfill outside of Rutland, Vermont. This solar farm also had an attached bank of utility size batteries. The interviews took place at Green Mountain Power's headquarters in Colchester and on the solar farm site.

THE BUSINESS OF ENERGY

Utilities in the United States are huge and very profitable businesses. As I started to understand how the electrical system worked in clear detail, I realized a book or documentary about clean energy integration would not do justice to the fullness of the story without including the perspective of at least one utility. They have the resources to affect large-scale change. They also have control over the power plants causing toxic waste and pollution. Without understanding how a utility works or how they are legally required to make business decisions, it is difficult for research to shed light into *how* utilities can transform to renewable energy electrical production.

Green Mountain Power is based in Colchester, Vermont, has 15 offices across Vermont, and 510 employees. It is the largest of 17 electric distribution companies in the state, with a customer base of about 265,000. GMP is the only investor-owned utility of these 17; the others are cooperatives and municipal electric companies.[2] In terms of national utility size, they are small. The top 10 utilities in the U.S. all serve customer bases of well over 2 million people. But their influence nationally has earned them top spots on national and regional rankings for everything from battery use and renewable energy to best customer service. For 4 years in a row, from 2017 to 2020, they won a spot on Fast Company's list of Most Innovative Companies in the World.[3] Mary Powell also received many awards for her leadership, including Executive of the Year from Utility Dive, a leading business journal reporting on trends in utilities.[4]

AMERICA'S MOST INNOVATIVE UTILITY

The first thing that I noticed that was different about how Mary Powell talked about Green Mountain Power was her basic description of what they do. She only partially talked about providing electricity and used the word "utility" simply because most people understand it. But she sees GMP as having a different conceptual and philosophical starting point, as well as a different business model. In her words:

Green Mountain Power . . . we are a very large utility for Vermont. We serve 80% of Vermonters. But really, how we view ourselves is an energy transformation company. . . . we are in the business of energy as a service, and we are in the business of helping Vermonters realize their dreams around energy. And, if you know anything about Vermonters, you know that the large percentage of them, their dreams around energy are really to reduce costs, to reduce carbon, and to be incredibly comfortable in their own homes and to have energy that's resilient. So we really see ourselves as an energy transformation company, not as an electric utility.

Research backs this up: Vermont has achieved significant results in climate and energy policy and practices around. In 2014, researchers at the Vermont Law School argued that their state is a national leader, and has "accomplished impressive feats across the domains of energy efficiency, renewable energy, the smart grid, and energy planning that deserve to be considered by, perhaps even replicated in other states and countries."[5] Looking broadly at several different utilities within the state, they cite Vermont's 2011 ranking as first in the nation for energy efficiency programs and an energy sector that is the cleanest and least fossil-fuel intensive in the entire nation.[6] They specifically reference Green Mountain Power, including a "smart meter" program that allows consumers to track and monitor their energy use. They also look at how Green Mountain Power, which is the largest utility in Vermont, is part of an ecosystem with other smaller municipal and cooperative utilities.

The World's First Certified B Corporation Utility

Researchers around the world are looking at the intersection of entrepreneurship and energy.[7] Research on social entrepreneurship looks into how theories of entrepreneurship integrate with other more common business practices.[8] While social entrepreneurship is perhaps often regarded as a somewhat new phenomenon, it has been a practice for over 20 years, with associated academic research and well-established practices of using business to drive positive social change.[9]

GMP is a perfect example of where several different identities intersect and how they foster progress for clean energy and surrounding communities. GMP is a Certified B Corporation, commonly referred to as a "B Corp." The official definition of a B Corp: "business(es) that meet the highest standards of verified social and environmental performance, public transparency, and legal accountability to balance profit and purpose. B Corps are accelerating a global culture shift to redefine success in business and build a more inclusive and sustainable economy."[10]

When she took the helm of Green Mountain Power, Mary Powell led it through the process to recertify as a B Corp, joining the ranks of many other industries with prominent social responsibility goals. Some of the more well-known B Corps are Ben & Jerry's, Patagonia, Eileen Fisher, and Kickstarter—which all sell products or create services far less solemn than electrical grids.

GMP is the first utility in the entire world to become a certified B Corp.[11] B Corps can have various areas of social focus, although many tend to have a strong focus on environmental, labor, and equity issues.

A typical corporation is legally required to make decisions of greatest benefit to their investors and to achieve the greatest profit, regardless of the other impacts of those decisions. A B Corp, on the other hand, has additional governing principles, and their corporate bylaws allow them, legally, to make their decisions based on other aspects than just profit. There are 3,500-plus B Corps today, in 70 countries and 150 industries, and they do everything from providing mortgages to selling loaves of bread. To become certified as a B Corp, a business must demonstrate verified impact in five areas: governance, workers, community, environment, and customers.[12] When a utility makes this choice, they are truly looking at transformation, starting in the boardroom—for example, by defining themselves not as a utility but as an energy transformation company.

Many of these technical and business innovations are available to any company that wants to implement them. However, it takes a dedicated person at the helm to guide a company to certification as a B Corp. While there are many definitions of and modalities for social entrepreneurship,[13,14,15] a certified B Corp is a specific business model.

I asked Mary why it was important to function as a benefit corporation or even a non-certified social entrepreneur business, and how it influenced what they did:

> We are a B corporation, which you probably have also heard, and I think we're still the only benefit corporation utility in the world. And I think it is the ultimate in sort of demonstrating to our customers and to ourselves that we are fundamentally social entrepreneurs. That we exist for the socioeconomic and environmental wellbeing of Vermont. That's why we fundamentally exist.

She elaborated on the internal process that a business goes through and how the certification was important to have; however, what mattered was the corporate culture that was behind the certification:

> The advantage of being a benefit corporation is intense. The interesting thing is, I believe, it was already our cultural foundation. By the time we actually went through the change in corporate bylaws and the certification process to become a benefit corporation, what was really exciting is that we already were. We already were a company that focused on energy transformation and cultural transformation. That said, being a part of a broader cultural community in business that's around bringing benefits, using energy in our case as a force for good, has been incredibly powerful.

I asked how being a B Corp affected the way they plan and run their operations:

> I heard the other day that some companies look at things in the next quarter. You always have to look at how you're going to do in the next quarter, but even more important is you have to be thinking about the next quarter-century. We look at the next quarter-century, and we think about how we drive down costs for Vermonters in the next quarter-century. There is absolutely no doubt in my mind that the work we're doing to create resilience has economic benefits for Vermonters.

GMP's Innovations

My first question was about GMP's technical and mechanical innovations which earned them the title of "most innovative" utility. Mary described their approach and what they did internally to transform the business culture:

I have long said that culture eats strategy. It was {Peter} Drucker who came up with the line . . . I've always added that culture eats strategy for breakfast, lunch, and dinner. One of the things that I have found in organizations, in communities, in serving customers is that focusing on the culture is the most important thing, particularly as it relates to an organization.[16]

At Green Mountain Power, a huge part of being focused on our customers, and being obsessed with our customers, and loving our customers started with changing our insides. How we operate. How we relate to each other, creating this one team which serves Vermonters. We know that Vermonters care a lot about the environment. We care a lot about the environment . . . we also care about clean energy and green energy. As a part of that, as I was transitioning to the role of CEO, we launched what was viewed (at the time) as a crazy ambition around clean energy. We launched an energy vision back in 2008 to provide low carbon, low cost, incredibly reliable power to Vermonters that was all about dramatically ramping up local renewable energy options.

This interview, in 2018, was after a decade as CEO. At the time, they had already achieved over 60 percent renewable energy in their energy mix (the mix of various types of electric generation, such as solar, wind power, hydropower from dams, and fossil fuel sources such as nuclear, natural gas, and coal):

At the time, back in 2008, I think we were somewhere around 20 or 30% renewable. We are now at 63% renewable.[17] *We're now 90% carbon-free in our portfolio. That energy vision—I call it "energy vision 1.0"—because we have moved on—that energy vision 1.0 was about dramatically leaning into the benefits of solar technology, really accelerating a consumer-led revolution to solar technologies in Vermont. It was also part of building local renewable energy options, like a wind farm for Vermonters at a much cheaper price than we ever could have "rented" wind energy from somebody else.*

I asked her to elaborate on what was meant by the energy vision stages:

Energy vision 2.0— and I can't wait for 3.0 because I know when you have an innovative, transformative, gritty culture, that you're ready to grab whatever that next technology is—our 2.0 is about how we embrace fundamentally, what is the greatest disruption happening in our industry . . . moving away from the "bulk power" system way of thinking and of delivering energy.

Elsewhere in this book, people involved in community transformation of energy also mentioned this disruption—the need to change the grid and change the production of electricity. Many stakeholders in energy fields recognize that large centralized power production, distributed over miles upon miles of power lines, is no longer working well. And that distributed production, such as rooftop solar or smaller microgrids, provides many benefits in terms of a secure energy supply, the ability to use renewable energy sources, and economic benefits.

The industry term for this—"distributed generation"—comes from distribution of the means of production throughout the area the power company serves, for example, on rooftops, rather than the traditional system of centralizing production in one large power plant.

Mary continues describing how and why GMP is disrupting this system:

> *The work we're doing is fundamentally reversing the system that we have today where there's this big, big bulk power delivery system that's sending energy one way into communities, homes, and businesses. To one where the primary system of energy creation is in those homes, businesses, and communities. And this big system becomes more of the backup system versus the primary system. Changing that around means fundamentally we will have a system that is not just more locally produced but is going to be inherently more reliable for Vermonters. We also believe that fundamentally the traditional, bulk system is just going up in cost every year. Now that new options are available, by leading and accelerating this adoption, we will also be able to save Vermonters literally tens of millions of dollars of what we otherwise would be paying if the bulk system is the primary system.*[18]

Some glimpses of what this means in a practical sense include GMP being the first U.S. utility to install Tesla Powerwall batteries into people's homes, starting in 2015.[19,20] The Powerwall made news when it was released as the first battery for home energy systems. It was priced within reach for many middle-class families; however, consumers could only access this new technology if their utility allowed them to install a battery. Unfortunately, some utilities cannot or will not accept the Powerwalls. These reasons can include technical ones like the need to test for fire safety in crowded urban environments. However, some utilities are also simply not ready to add batteries.

Since they started allowing these batteries, GMP has expanded their offering beyond the Powerwalls, allowing customers to buy other models of batteries, thus prioritizing customer choice.[21] This was one of the decisions which earned them the "Most Innovative Utility" status. Another part was that once many of these individual batteries were installed, they wove them into a bigger system and used the home batteries as a backup for the main grid, leading to the distributed generation system that Mary described.[22]

Another innovation was mixing different kinds of renewable energy, which can help address intermittent power and stability issues described in other chapters of this book. Mary describes how they started building wind farms, which can create a stronger power system because solar and wind tend to produce the largest amount of power at different times:

> At Green Mountain Power, part of our energy vision 1.0, where we looked at developing renewable resources like building the Kingdom Community Wind project. One of the things we really liked is the counterbalance of having solar and wind available for us and for our customers. Wind energy tends to produce best in the winter months, generally speaking. It's when you have the windy months. Solar, as you would guess, tends to produce best in the April through August timeframe. We felt as renewable resources, while they both operate year-round, in terms of their greatest value, they were complementary to each other.

This intermittent nature of renewable energy is a major hurdle for utilities accustomed to fossil fuel power production, which works by burning gas, oil, or coal, and thus can be turned on and off more or less as needed. Intermittency, often expressed as "the wind doesn't always blow, and the sun doesn't shine," is part of the nature of these energy sources. Some utilities use intermittency as a reason not to pursue renewable energy at all. Others see the inherent value in solar and wind power and are designing their energy systems to work with intermittency. The innovations GMP introduced have allowed them to plan this energy input into their system.

Energy Storage—More Than Just a Battery

An exciting point of progress for all utilities is the development of ways to store this energy. For example, if the sun shines most from 11 a.m. to 3 p.m., storage would hold onto that electricity until 6 p.m. when demand climbs because people come home from work. "Storage" is, in layperson's terms, simply a battery such as the Powerwall and other brands. Batteries allow energy companies to finally have an affordable reservoir to hold energy produced at noon and use it at midnight. Chapter 2 addressed the immediacy of electricity and how the inability to store it was one of the technical hurdles that had, for many years, made solar and wind power more difficult to use. Now, as batteries come into consumer markets, it represents a potential for changing structures, designs, and means of using electricity on a scale of the change that cell phones were to landline phones.

Many brands of utility-scale batteries are on the market, with home-sized ones becoming affordable in the last five years. These batteries are produced in varying sizes, some for completely off-grid use. For example, Blue Planet Energy, based in Hawaii, has made utility-sized batteries designed to withstand rugged rural conditions. (See chapter 8, on La Riviera, for how Blue Planet Energy's batteries were used to create solar-powered water pumps in Puerto Rico.)

Mary continues to describe how they implemented energy vision 1.0:

We fast-followed with storage. Again, as you may know, we were the first utility in the country, I think actually maybe in the world, to work with Tesla on delivering home storage solutions,[23] *which is a perfect complement to home solar solutions and community storage solutions, and a perfect complement to community solar solutions. Because when those things work together, you avoid some of the energy system disruptions that some of the more traditional utilities have talked about as a way of, in my mind, sort of resisting solar technology. Solar technology by itself would literally just push the energy load of what we're all using later into the day. When you combine it with storage technologies, that is when you really start to create a powerful change around moving towards a community, home, and business-based system. We're never thinking of just that next technology. We're thinking of what's next after that. And, so we really led the consumer-led revolution in the U.S. around solar energy way back in 2008.*[24]

Vermont is no stranger to intense weather, and has its share of blizzards, ice storms, torrential rain, and heat waves. It is also an extremely mountainous state. As a result, people tend to cluster in communities located between long stretches of mountains, instead of living in sprawling suburbs, with evenly and continuously distributed population, or urban areas, with densely packed population. This means that Vermont's power lines need to go through mountain ranges, making them subject to intense winds, snowfall, ice storms, and other types of inclement weather.

A final aspect of the redesign that Mary discussed was changing how to address this, and how GMP, rather than try to protect all of these lines against inevitable weather, decided to redesign the system:

One of the things that informed my thinking about this was seeing what happens to the poles and the wires and pretty significant infrastructure when Mother Nature rolls in, with much greater force than we were used to seeing happen on a regular basis. And what happens is that these big poles and these big wires . . . They get knocked down, and they get put into little balls of wire. And, it is a great example of how that type of approach fundamentally is ripe for transformation—from an economic perspective, from an environmental perspective, and from a physical perspective."

As with Puerto Rico and New York, storms had done significant damage. Damaged lines mean that GMP had to repair them constantly, which was costly and difficult to keep up with Mother Nature:

Storms and climate change have very much accelerated and influenced our thinking about this community home and business-based energy system. I've lived through a ton of once-in-100-year-events, once-in-200-year-events. And there's not an event like that when I have not put on boots and a hard hat and been out visiting our folks that are restoring power. Because it's really hard work. The least I can do is support them. So I get out there and I support them.

The big "aha" for me was when I first started doing that, and I saw projects that we had just finished, like an amazing, you know, strengthen and harden the lines to make them more resilient for weather events. And those same projects become—I coined the term "twigs and twine"—because they become twigs and twine when Mother Nature moves in. In fact, the Hydro-Quebec transmission system[25]—if you look at a picture of it, it is massive—was a steel structure that became twigs and twine when the ice storm of 1998 hit. There is no real true resilience when you think about it. When you think about this, it is an economically inefficient way to deliver energy to begin with. And that's part of why it's getting more economically inefficient because we're trying to make that twigs and twine as strong as we can, to be as resilient as we can, and to be as reliable as we can. But now there are devices that are saying to us as consumers "we don't have to use that system as much anymore."

Now that it is technically possible to build a grid to be less conducive to becoming twigs and twine, I asked whether that would save customers money. Mary replied:

Well, whenever a company, Green Mountain Power, or any company, needs to rebuild the infrastructure serving its customers, the customers, of course, pay for that. Because there is no reason for that infrastructure to exist except for the customers. Every single thing we do at Green Mountain Power, as we think about energy transformation, as we think about resilience, as we think about going from a bulk twigs and twine system to a distributed system, is all about the math of creating a more economic system for Vermonters.

For more on the cost of rebuilding the grid, see the chapters on Isabella and Salinas, and chapter 7, on Isabela. In particular, refer to the quote from Governor Mapp of the U.S. Virgin Islands, who testified to Congress after Hurricane Maria that they rebuilt their grid five times after hurricanes, and they would like to design it in a way to survive storms rather than have to rebuild yet again.

Redesigned Solar and Wind Projects

Two examples of power production on this redesigned, distributed-generation grid are:

1) Mary suggested I take a trip to one of their solar installations outside of Rutland. She described it as a representative example of how they want to continue to innovate. The solar power system is stationed on a former landfill very close to a high school. Citing it on a landfill repurposes land that cannot be used otherwise and that preserves arable lands, wild areas, pastures, and forests.

Several utility-scale batteries were arrayed in a backup system at the Rutland site, and charged by the solar panels and the grid. Vermont clearly does not receive the same amount of sunshine as Puerto Rico or Nevada, but anyone living in a colder state who believes they do not get enough sun for solar to work should consider that Vermont and other neighboring New England states frequently top the lists for most use of solar. Vermont certainly gets sufficient sun for solar to be widespread; however, with fewer hours of sun per year, the battery systems become more important.

The engineers I spoke with also explained that a microgrid attaches the entire system of solar panels and attached batteries to the nearby high school, which was within site distance from the landfill/solar farm. This design meant that in the event of a storm, with the grid becoming twigs and wine and causing a power outage, the high school could island off. The solar and battery system make the school an emergency center for blackouts.

2) Mary also described the Kingdom Wind Farm in Lowell, which generated a certain amount of controversy in the media when built, because of its location which was high on a ridge, made it quite visible. However, the community voted not once but twice about whether they wanted the wind farm. Each time they overwhelmingly voted to keep it.

The wind farm was built and is owned by Green Mountain Power. Some wind farms are built and operated by outside companies and sell their power to the utility. When the wind farm was proposed, GMP executives encouraged the town council to have a public vote and were adamant that they would not build the wind farm if the town voted it down. The wind farm passed with a 75 percent vote the first time. The council felt they needed to vote again a year later. The wind farm won—by an even higher margin this time. The town receives revenue from the wind farm, which is applied to reduce residents' property taxes, helping residents benefit from the wind farm's presence.[26]

Public Reception and Opposition

As we discussed the customer-centered approach, I wondered how GMP managed the inevitable conflicts around clean energy transformation. For example, the controversy that greeted the Kingdom Wind Farm as well as controversies about solar projects. I asked Mary how they responded to these conflicts and controversies:

One of the things that has been important to us is community support. When I say customer obsession, we are doing what 80% of our customers say they want. That means there are 20% that maybe do not want something. I think it's important to approach things from the perspective of understanding "what do the vast majority of your customers want?" What do the vast majority of your communities want? And then to try to do it in a way that incorporates those who have differing views. Vermont has transformed a lot in energy over the last five or six years. And, whether it is energy transformation or shopping malls, you are going to have people that do not want to see a change. I think the most important thing is working with Vermonters and working with those who have differing views, to the best of your ability.

Internal Cultural Changes

As mentioned above, Mary initiated the process to certify GMP as a B Corporation. On their way to becoming officially certified, she discusses how she focused on internal corporate culture as the heart of these other changes. I asked what that meant:

Well, first of all, I turned down the opportunity to join Green Mountain Power a number of times because I didn't have a desire to work in what I viewed as an organization full of smart, nice people, but an organization that felt like a traditional, old-fashioned, bureaucratic, analytical, slow-moving organization. So, I said no a couple of times. Obviously, eventually, I said yes because I felt like there was a reason it kept coming up. I felt like, on some level, that I was supposed to come to this organization. As it turned out, initially the reason was to completely and radically transform the whole culture. And so, going back, I was leading what was then, and still in some parts of the country, a radical transformation away from a very bureaucratic way of thinking about the business, to one where we are obsessed with customers, we love customers, we disrupted our whole infrastructure to serve Vermonters.

She continued, describing changes in workspaces, the basic social environment in which employees interact, and how the company deliberately creates culture:

Our transformation was on how we actually behaved, and it was around, again, being customer-obsessed. We took very traditional utility stats, like a 60%

> customer satisfaction level. We now have a 94% customer satisfaction level. We have 92% customer trust. And that is because we made a very lean, non-bureaucratic, open, fast-moving organizational structure.
>
> I'm in the middle of a colorful, energetic space where the linemen arrive in the morning, right next to where I work. Other folks in leadership and engineers are all clustered in the same area. So that's really how the transformation started. I'm a firm believer in the line "culture eats strategy" because I firmly believe that the most important thing organizations need to do is focus on the culture of the team. The messages that are being sent every single day about priorities. Not through posters on walls and newsletters, but in terms of how people actually behave.
>
> Are we working as one team? Am I working as much with the frontline of the organization as I am with the leadership team? Are we sending messages that the teams that do direct service to customers are the ones that we value the most? Are we sending messages that customers are the most important people in our lives? Those fundamental qualities then help you drive whatever business strategy it is that you need to drive. You can't innovate and transform for Vermonters if you're not willing to innovate and transform for yourselves within your own organization. So I am a firm believer that culture is the most important thing to focus on.

Given this belief in cultural transformation, and the transformation of the energy system, I asked her what her thoughts were on the clean energy revolution and whether she believed GMP is on the front lines of it:

> I absolutely believe there's a clean energy revolution. I believe, more than anything, it is finally a technological revolution. It is funny because when I joined the company, one of the first things I said was, "this system has been around forever." Something has got to come along to disrupt this. In every other space, whether it is media, it's the telephone, whether it is how we receive entertainment, even transportation . . . there has been a radical transformation. So what is it in the utility business?
>
> I have long felt like this is an industry ripe for technological disruption, and then you combine that ripeness for technological disruption with a consumer desire and concern around clean energy. It is a perfect opportunity for radical transformation.
>
> I love that line: "If you sit on the lid of progress, be prepared to be blown to bits."[27] We have never been about sitting on the lid of progress. We are all about accelerating progress in a way that can be transformative.

At the end of 2019, after over a decade of leading GMP, Mary Powell stepped down. The senior vice president took on the role of CEO to carry on the same initiatives. Her influence about innovation and cultural change lives on—GMP has new initiatives such as offering special rates and other incentives for people who buy electric cars, which keeps the cost of charging a car battery lower than filling a tank with gasoline.

Perhaps more important than any one specific technological innovation is her introduction of how to change the culture and her suggestion to the industry that energy transformation is about people as well as equipment. Asked about her legacy, she said:

> "*I feel like what my legacy will be—maybe it's what I hope it will be—is that all the great stuff came out of my love of people." I feel the biggest impact I made is the complete transformation of the culture. . . . We created a culture of the elimination of bureaucracy. Of moving with trust and respect for each other. I hope that's the cornerstone of my legacy, my leadership style, and my orientation towards life.*[28]

CONCLUSION

Every aspect of the structure of a utility—the business model, core values, internal culture, specific bylaws, and leadership—is profoundly influential in how utilities create power. Green Mountain Power has set a standard for thinking miles down the road, incorporating clean energy technology into electricity production and delivery, building new infrastructure, creating new business models, and otherwise evolving every aspect of its business model.

Cities, states, and utilities across the country which are looking to adopt clean power would be well-advised to look into GMP's examples for their own transformation.

NOTES

1. Department of Public Service, State of Vermont. "Vermont Electric Utilities." *Vermont Official State* Website. https://publicservice.vermont.gov/electric

2. VELCO. "Who's Who in Vermont's Electric System." Vermont Electric Power Company. https://www.velco.com/about/learning-center/vermonts-electric-system

3. Fast Company. "Green Mountain Power: Most Innovative Company | Fast Company." *Fast Company*. 2020. https://www.fastcompany.com/company/green-mountain-power.

4. Gheorghiu, Iulia. "Executive of the Year: Mary Powell, Green Mountain Power." *Utility Dive*. 2019. https://www.utilitydive.com/news/ceo-mary-powell-vermont-green-mountain-power-dive-awards/566247/.

5. Sovacool, Benjamin, Alex Gilbert, and Brian Thomson. "Innovations in Energy and Climate Policy: Lessons from Vermont." *Pace Environmental Law Review* 31, no. 3 (2014a), 651. https://digitalcommons.pace.edu/pelr/vol31/iss3/2/.

6. Sovacool, et al. "Innovations in energy and climate policy."

7. Kummitha, Rama Krishna Reddy. "Social Entrepreneurship, Energy and Urban Innovations." In *Mainstreaming Climate Co-Benefits in Indian Cities*, pp. 265–283. Springer, Singapore, 2018. https://link.springer.com/chapter/10.1007%2F978-981-10-5816-5_11

8. Shaw, Eleanor, and Sara Carter. "Social Entrepreneurship: Theoretical Antecedents and Empirical Analysis of Entrepreneurial Processes and Outcomes." *Journal of Small Business and Enterprise Development* (2007). doi: 10.1108/14626000710773529/full/html

9. Dees, J. Gregory. "The Meaning of Social Entrepreneurship." 1998. http://www.redalmarza.cl/ing/pdf/TheMeaningofsocialEntrepreneurship.pdf

10. B-Labs. "About B-Corps." *B-corporation website*, accessed January 2, 2021. https://bcorporation.net/about-b-corps

11. Powell, Mary. "How to Power a City." Interview by Melanie La Rosa. October, 2018.

12. B-Labs. "About B-Corps." *B-corporation website*, accessed January 2, 2021. https://bcorporation.net/about-b-corps

13. Ashoka. "Social Entrepreneurship." Ashoka.org, accessed January 2, 2021. https://www.ashoka.org/en-us/focus/social-entrepreneurship

14. Dees, J. Gregory. "The Meaning of Social Entrepreneurship." 1998. http://www.redalmarza.cl/ing/pdf/TheMeaningofsocialEntrepreneurship.pdf

15. Guzmán, Alexander, and Trujillo, Maria-Andrea. "Social Entrepreneurship–Literature Review (Emprendimiento Social–Revisión de la Literatura) (Spanish)." *Estudios gerenciales* 24, no. 109 (2008): 109–219. https://papers.ssrn.com/sol3/papers.cfm?abstract_id=1756171

16. Powell, Mary. "How to Power a City." Interview by Melanie La Rosa. October, 2018.

17. Green Mountain Power. "Energy Mix." *Green Mountain Power website*, accessed January 2, 2021. https://greenmountainpower.com/energy-mix/.

18. Powell, Mary. "How to Power a City." Interview by Melanie La Rosa. October, 2018.

19. T&D World Magazine. "Green Mountain Power Files Plans to Offer Tesla Powerwalls to Customers." *T&D World Magazine* website, December 8, 2015. https://www.tdworld.com/distributed-energy-resources/article/20965959/green-mountain-power-files-plans-to-offer-tesla-powerwalls-to-customers

20. Green Mountain Power. "Tesla Powerwall." *Green Mountain Power* website, accessed January 2, 2021. https://greenmountainpower.com/rebates-programs/home-energy-storage/powerwall/

21. Gheorghiu, Iulia. "Executive of the Year: Mary Powell, Green Mountain Power." *Utility Dive*. 2019. https://www.utilitydive.com/news/ceo-mary-powell-vermont-green-mountain-power-dive-awards/566247/.

22. Fast Company. "Green Mountain Power: Most Innovative Company | Fast Company." *Fast Company*. 2020. https://www.fastcompany.com/company/green-mountain-power.

23. Tesla, "Exclusive Green Mountain Power Solar and Powerwall Offer." *Tesla.com*, accessed January 2, 2021. https://www.tesla.com/gmp-bundle

24. Dostis, Robert. "Presentation: House Energy & Technology: Green Mountain Power Corporation" *Vermont Legislature* website, January 19, 2017.

25. HydroQuébec. "From the Power Station to Your Home." *HydroQuébec.com* website, accessed January 2, 2021. http://www.hydroquebec.com/learning/transport/parcours.html

26. Valley News. "Lowell, Vt., Supports GMP Wind Power Project — Again." *Vnews.com*, March 5, 2015. https://www.vnews.com/Archives/2014/03/a13WireVtTM-ls-vn-030514

27. California Museum. "Biography of Henry J. Kaiser." *Californiamuseum.org*, accessed on January 2, 2021 https://www.californiamuseum.org/inductee/henry-j-kaiser

28. Marcel, Joyce. "She was Fast, Fun and Effective: Mary Powell leaves GMP." *Vermont Business Magazine*. January 18, 2020. https://vermontbiz.com/news/2020/january/18/she-was-fast-fun-and-effective-mary-powell-leaves-gmp

Chapter 6

Salinas, Puerto Rico
Solar for Survival

INTRODUCTION

When Hurricane Maria devastated Puerto Rico on September 20, 2017,[1] residents of Salinas, Guayama, and other communities on the southeast coast already knew the best direction for transforming their energy supply back to functionality. With two massive power plants polluting the nearby Jobos Bay, they advocated for years for solar power and other alternatives for centrally produced, fossil fuel power generation. Across the island, the excruciatingly slow rebuilding of the grid left the entire population of 3.4 million people struggling without electricity for months. For some, it was a full year before they had electricity again. Along the way, there were numerous additional blackouts and transformer explosions. People waited, without electricity or solid plans to restore it, while the government entered into contracts initially with private companies from the mainland, like Whitefish and Cobra, and also with the public power utility in Puerto Rico, the Puerto Rico Electric Power Authority (PREPA). This record-setting wait to restore the grid, particularly for people living in rural or remote areas, revealed the deep flaws in the design and construction of a long-distance power grid on a mountainous island frequented by hurricanes. It also revealed the deep flaws in the management and maintenance of a power system that was widely considered dilapidated and sub-par long before the hurricane leveled it.

CHAPTER SUMMARY

This chapter explores, through the experiences of a far-sighted group of people in southeastern Puerto Rico, the considerable impact that rooftop solar and other types of distributed generation have on the ability of a power system to withstand hurricanes and other devastating natural events. This chapter also looks equally at the meaningful difference this makes in the lives of community members, for their economic well-being and their health.

It is well-accepted that solar, wind, and many types of renewable energy create electricity without creating pollution. Less understood is their role in creating reliable electrical systems, surviving storms, and providing additional benefits beyond electricity. In the wake of one of the worst blackouts on record, widely regarded as being induced by climate change, how can energy transformation realize these benefits? How can we transform energy systems?

BACKGROUND

When Hurricane Maria struck Puerto Rico, I was in Nevada, filming with executives from NV Energy. We had all received news alerts, but only one day after this massive travesty, reliable and detailed information was sparse. The headlines spoke to widespread devastation, announcing that the entire power grid was down. But nothing explained how a hurricane, regardless of size, category, or any other factor, could knock out power for a population the size of Connecticut's. I wanted to understand the mechanics of how this could happen.

Since I was filming with executives from a utility, during a pause, I took the opportunity to ask them their perspective: how could a hurricane shut down an entire island? Sections of it, of course. A few cities, perhaps. But an entire island? They also seemed like they were trying to understand as well. This was the first day after the hurricane, and they shared that, from what they had read, it was about the lines going down and not the power production itself. The natural gas, oil, and coal-burning power plants were functional. Over the majority of the island, renewable energy sources were functional, except those very close to where the hurricane made a landfall on the eastern coast, where a wind farm was destroyed and a solar farm was seriously damaged. The blackout, however, was caused by the failure of the transmission and distribution systems that delivered electricity to people's homes.

For many, these two pieces of the system—power production and power distribution—might seem inseparable. But they are, in fact, two very separate pieces, with two different purposes. Understanding this difference and looking closely at how these two systems operate reveals volumes about the role of solar for survival.

When I started researching for people involved with solar development in Puerto Rico, it did not take long to find Ruth Santiago, a lawyer and environmental policy expert with a long history of fighting the toxic waste in the Caribbean. Her work with environmental groups including, Comite Dialogo Ambiental, Inc. ("Dialogo," Environmental Dialogue, Inc.), Iniciativa De Ecodesarrollo De Bahia De Jobos[2] ("IDEBAJO," Initiative for the Eco-Development of the Jobos Bay), as well as with the national organization, Earthjustice,[3] had earned her wide recognition as a voice protecting public health and advocating for clean air and water. Ruth is part of a wide network of community residents, activists, youth, researchers, engineers, and others whose combined efforts have created a strong local public education program on environmental issues, particularly related to the Jobos Bay watershed and promoting holistic management of the area's natural resources. IDEBAJO's programs also included education on solar energy, which they had already been doing for years. The Junta Comunitaria de El Coqui—the El Coqui Community Board—part of IDEBAJO, started a solar energy proposal in 2014—three full years before Hurricane Maria.

These programs had covered significant ground already, and when Hurricane Maria hit, the community already understood off-grid solar technology. In fact, IDEBAJO had already been organizing to be able to build rooftop solar power systems and had formed a working group that included Ruth, researchers from the University of Puerto Rico, Mayagüez, including Dr. Efraín O'Neill Carrillo, Professor at the Department of Electrical and Computer Engineering, University of Puerto Rico Mayagüez, a leading researcher on energy issues; and community activist Roberto Thomas, Coordinator, IDEBAJO, who is involved with community-based initiatives throughout the region. In addition, several local residents attended, along with about a dozen students from high schools and the University of Mayagüez. They were members of "Coquí Solar," named after the coquí (pronounced ko-kee), a small frog whose signature loud chirp is heard throughout the island. The coqui is an unofficial and widely beloved national symbol. ("El Coquí" is a common name seen throughout Puerto Rico. In the case of Salinas, it is also the name of an "urbanización," or neighborhood, in which there is a community center, Centro Communitario El Coquí.)

During our initial phone conversation, Ruth told me about the work of Coquí Solar, including their focus on community solar. Throughout Puerto Rico, at this time, families were embracing rooftop solar, which had been available on the island for some time. Businesses and others who could afford to were expanding industrial-scale solar farms. Everyone was learning about new technology like off-grid batteries. "Community solar" or solar-powered communities, however, were an even newer approach. In looking at how solar could help people survive hurricanes, it is also important to look at how energy economics are critical to that survival, along with the overall design of a solar power system, its durability, and how it integrates into people's daily lives.

To address the question about what type of a system was in place that could result in a power outage for an entire island, the answer must look at how people address the flaws in the economic systems as well as the technical systems. At the time of Hurricane Maria, Puerto Rico had the second-highest electric rate in the nation, with only Hawaii having more expensive power. There are different definitions of community solar—a detailed discussion of what it means in this chapter is below. Chapter 9, on Highland Park, and chapter 4, on New York City, both describe slightly different versions of community solar. One of the most important aspects; differentiating community solar from other types of solar power systems is that it is a system that lowers rates and benefits the community in ways beyond just providing electricity. Some variations of community solar provide shared ownership of the actual solar panels. Other approaches to community solar do not require ownership by the community of the actual solar panels. Instead, the panels and other equipment are collectively rented or bought by a larger community group, allowing people to take advantage of clean power without the expense of purchasing, installing, and maintaining the equipment. For example, one of the things Coquí Solar is proposing is that PREPA install solar systems on rooftops and interconnect those systems at the community level. This option would accelerate the necessary transformation of the electric system in Puerto Rico and create a more reliable source of power for all Puerto Ricans.

METHODOLOGY

The material for this chapter was recorded during two filming sessions: the first, in March 2018, which was six months after Hurricane Maria, during a community meeting on solar taking place at the El Coquí Community Center in Salinas. The second, in October 2018, was a follow-up interview done just a bit over one year after the hurricane. Finally, additional recording took place during the first island-wide Transformación Energética Desde Las Comunidades, or Community Energy Transformation Gathering, held at Interamerican Law School in San Juan in December 2018.[4] As with all of the content in this book, I used narrative inquiry to gather the information.

THE TRANSFORMATION: COQUÍ SOLAR

On a blazing Saturday morning, the El Coquí Community Center was crowded. It was six months after Hurricane Maria. The start of the next hurricane season loomed a mere 3 months away. Inside the community center, relief supplies—generators, clothing, cases of water—were piled up along a wall. About 20 people had gathered for a meeting on community solar.

In our first interview, Ruth summed up their challenge:

I think most people's dream is to actually have some type of solar equipment on their rooftops, the problem is being able to afford it, right? How are we able to finance that? Especially for the poorest communities, the communities that have been most excluded and marginalized, and that are the environmental justice communities in terms of being impacted by that dirty fossil fuel generation. And yet, being in the shadow of these plants, so to speak, and not being able to afford an alternative . . . that's where we're looking to promote—that alternative.[5]

During the meeting, the focus of the discussion was to ensure the most widespread community benefits during the rebuilding. Relief and humanitarian efforts brought solar panels, racks, batteries, and installers to the island in significantly expanded quantities. Many communities received small emergency solar power systems, similar to those distributed by Solar Libre in Isabela, discussed in chapter 7. But, the members of Coquí Solar wanted to plan how they could use this influx of solar to support vulnerable community members. How could resources be directed to first support the elderly and bedridden, then extend to the wider community? How could this influx provide jobs and help people rebuild, or build, their lives? How could solar help families save money in the long term? In short, how could the post-hurricane solar influx provide more than just electricity?

Solar Power Systems

Throughout the United States, in localities with laws friendly to clean energy, getting residential rooftop solar has become relatively simple; a homeowner contracts with a solar company, a representative assesses the house and sizes the solar array based on the available sun and average electrical use. The size of the system is typically determined from the annual average use based on the individual's electric bills. The homeowner finances the equipment as they would any large purchase, like a car. Sometimes they finance it through the solar company or another entity. Sometimes they take out a loan from a credit union or bank. Sometimes they purchase it outright. The vast majority of these systems are "grid-attached" or "grid-tied." The solar power they generate feeds into the main power system, and the family draws back from that system while receiving a credit on their power bill. To be "off-grid" or have a stand-alone system requires a very large battery to store the solar power, as well as laws that permit this. Until recently, off-grid solar power systems have been technically challenging and financially out of reach, which is one reason that they are not permitted in some areas. The introduction of consumer-affordable batteries, or "storage systems" as they are sometimes referred, is a game-changer on all levels: technical, financial, and policy.

At this time, six months after the hurricane, according to Bloomberg Law, residential solar sales in Puerto Rico had nearly doubled from before the hurricane.[6] Some of this development was for grid-attached systems. But communities and businesses—those who could afford them—purchased large, off-grid batteries, effectively creating self-sustainable, resilient communities that could detach from the grid. For reasons of resilience and economics, most of these communities continued to be connected to the grid, even if they had battery-plus-solar systems, but the option to detach is now within reach, which was important given the notoriously inconsistent service delivered by Puerto Rico Electric Power Authority, or PREPA.

This desire for solar was widespread. Coqui Solar and Queremos Sol, an island-wide collaboration of groups working for solar on the island, promote rooftop, onsite solar, and other alternatives as opposed to land-based solar arrays for many reasons. As mentioned above, they locate the power production at the site of its use, reducing the need for transmission wires. Rooftop solar also preserves habitats and agricultural land and avoids impacts to ecological areas. The Queremos Sol proposal (detailed below) would allow lower-income families the same access to solar that wealthier families already had. For example, one of the solar companies I spoke with was Yarotek. During an early 2018 research interview, executives from Yarotek shared with me that they were working with a resort community, Palmas Del Mar, to create an off-grid battery backup system (LaRosa, field notes, 2018). Palmas Del Mar, like El Coqui, is a residential community located in southeastern Puerto Rico, although on the opposite end of the income spectrum. While these two communities have vastly different resources, they share the same sentiment of simply being tired of "the poor, unreliable, and very expensive electric service" and the feeling that "unfortunately, partisan politics and corruption have guided the decision making processes in PREPA . . . . we are now paying the price of corruption."[7] Palmas Del Mar residents, however, had the means to access solar nearly immediately. This provides a proof of concept that solar communities can work. However, it also highlights the need to ensure that solar communities, and the resilience, safety, and cost-savings they can provide, are available to everyone.

One of the design flaws that Hurricane Maria revealed for Puerto Ricans was that solar panels without a battery system—"grid-tied" systems—are of no use if the grid goes down. As Ruth put it:

People who did PV (photovoltaic) installations that were grid-tied, and didn't have batteries . . . after the hurricane, because the PREPA grid was not working, because all the transmission and distribution lines were down, they were also not able to generate energy, even if they had solar panels, because they had no battery system and they were connected to the PREPA grid. So definitely it should be done differently so that people have some battery systems, enough as needed to keep their basic appliances going, their lights, et cetera. I think that it can be done in a way that people are still connected to the grid if they need more energy, but once that main grid is not functioning, they can rely on their own backup systems.[8]

The same design issue was summed up, from a different perspective, by a Yarotek executive who explained it in shorthand: *"if the system is broken anywhere, it is broken everywhere"* (LaRosa, field notes, 2018). The same executive confirmed that their large solar installations (they had built several large solar fields throughout the island prior to the hurricane) were all working in the days immediately following Hurricane Maria. The damage to their systems was minimal. However, because the entire grid went down, nobody on the island could benefit from the power produced by these solar panels. Only by installing microgrids and batteries could they change the grid to address this fundamental flaw.

While becoming more affordable, solar arrays are still out of reach financially for a great number of people. Attaching a battery backup that can withstand the elements is even more costly. Six months after Hurricane Maria (and Hurricane Irma, two weeks prior), this set the stage for the danger that a reliable connection to modern life could be available only to the wealthy.

Empowerment

In areas where hurricanes occur frequently, another dire threat emerges, which has been explored in research on environmental communications and how underpinnings of language and perception help construct dysfunctional systems. There is a sense that energy systems never improve. That they are immutable, they have always existed as they are now, and that we cannot change these systems. This sense of futility is supported by the ways we perceive our energy systems. By introducing new ways of describing and perceiving the relationship of humans to their electric power, we can also begin to address this sense of futility and find ways that communities can—and do—affect their local electrical power system.

Environmental and communications researchers have explored this fundamental relationship with energy at conceptual levels. One such concept is energy coloniality, which is "constituted by a discourse and system that colonizes places and peoples to control different energy forms, ranging from humans to hydrocarbons" (de Onís, 2019). Adding to this, de Onís expands on another concept, that of "the interrelated 'energy privilege' . . . the study of privilege as essential for uncovering and resisting the domination of different energy forms, which fuels and is fueled by unjust power relations that benefit some and harm others." These are necessary concepts to understand the importance of community solar as a solution. Systems rooted in colonialism and privilege command the energy choices available to an entire population yet make choices that do not benefit that population. Smith-Nonini found that by looking at the social and health impacts through the lens of the debt crisis and colonial past revealed "insights into patterns of manipulation by politicians, bankers, and fossil-fuel companies that led to excess borrowing by the Puerto Rican public electrical utility, contributing to a neglect of infrastructure and high rates for citizens." Building on these concepts, one might think that the PREPA decisions were made to favor the wealthy and privileged. However, if the energy systems that existed at the time of Hurricane Maria were not even serving relatively wealthy and privileged island residents, like those in the Las Palmas, then who is this system serving? When people in Las Palmas bridled equally at the corruption and dubious service along with El Coquí residents, it is difficult to see who benefited from PREPA's energy choices?

If we understand the established systems as a series of choices made by multiple people in positions of authority, then we can see that the power system is indeed mutable. Suppose we see that using one form of electrical energy over all other forms, results from the choices made by a small handful of people. In that case, we can see that change is possible by either changing those decisions or changing that small handful of people. In line with this, Queremos Sol also proposes a transformation of PREPA's governance. However, if we also look through the lens of energy coloniality and privilege, when people with and without means are equally frustrated and demanding change, then who or what is being served? Certainly not the customers who are paying their power bills each month. The only stakeholders who want to continue in this system are the fossil fuel companies and related interests. And, when the mechanics of that system— the wires and poles—lay strewn, useless, in gutters and streets throughout the island, exposing inherent flaws in the design, why rebuild as it was before if the customers paying into the system are not served by it?

Journalist, author, and filmmaker Naomi Klein introduced the term "disaster capitalism" to describe the phenomenon of various types of disasters, including Hurricane Katrina in New Orleans, causing a second shock wave as people lost their homes and land, leaving entire cities open for corporate purchasing and redevelopment.[9] In the aftermath of Hurricane Maria, the imminent and very real threat of disaster capitalism loomed. The most notable evidence of this was the widely reported and highly controversial multimillion-dollar deals with two companies, Whitefish and Cobra.[10]

Scholars like Dr. Catalina de Onís cautioned about the probability of "energy colonialism" and advocacy for decentralized solar infrastructure to allow more choices and more public participation, particularly in light of the corruption of existing power authorities. Dozens of global solar companies rushed to support Puerto Rico in rebuilding with solar. They brought literal tons of much needed equipment to the island, donating significant amounts of it. This leveled two of the most universal and formidable barriers to clean energy transition on an island: the cost of equipment and shipping to an island. Along with these solar companies came brigades of volunteer linemen, solar installers, engineers, and other clean energy technicians. But with all of this beneficial investment and support, other companies joined as well, some with questionable motives and track records, many seeking their own multimillion-dollar contract. In this scenario, community solar is a powerful solution, with the potential not only to avoid the dangers of disaster capitalism and energy colonialism. It also has the potential to address many of the more deeply entrenched issues Puerto Ricans had been coping with: the debt crisis, lack of career opportunities for young people, school closures, and the many people leaving the island for better opportunities elsewhere (Hinojosa, J., 2018,[11] a Katz, 2019[12]).

This last factor—people leaving—is profoundly influential for the island's future. In fall 2019, the U.S. Census reported data from two studies showing that the poverty rate in Puerto Rico was 44.4 percent, more than triple the national U.S. average of 13.1 percent. A third of more people migrated from Puerto Rico to the U.S. mainland in 2018, reducing the number of island residents by 4.4 percent.[13] A 2018 study by the Pew Research Center found a similar number of people leaving, reporting that the island's population declined by 3.9 percent from 2017 to 2018, representing the largest year-to-year drop since 1950, the first year for which annual data is available.[14]

Two years later, a 2020 study using credit card data found that "out-migration continues to remain elevated."[15] Commonly referred to as "the exodus," many researchers document the people leaving the island; where they go, how often they return for visits or for good, the ongoing fluctuations in annual rates of out-migration, and the impact on the island's economy. But what about those who want to stay? What is their involvement in rebuilding as they see fit? What are the opportunities for them? What mechanisms are in place which support them to stay on the island? And how is the electric system and other essential services—or lack thereof—contributing to this exodus?

A well-designed solar power system would focus on the community it serves and be built in a way to address the technical limitations and design flaws that caused all of the wires and poles to end up strewn across the island. Another issue is that the influx of solar donations might have brought valuable equipment to the island, but technology alone does not elevate the skills and knowledge of island residents about using that technology and maintaining the systems. So when the next hurricane hits, and the volunteer solar installers are back home in Miami, South Carolina, New York, and other states, who will fix the systems?

A case in point, the influx of equipment from Tesla, which received significant media hype in the months following Hurricane Maria, turned into what one island resident called "a cautionary tale." Tesla pitched to the Puerto Rican government that they would install a microgrid on the smaller islands of Vieques and Culebra, located just off the eastern coast of the main island. They were given permission to do so and installed their cutting-edge batteries and solar arrays on hospitals, senior centers, and other buildings. The result? While many of the Tesla donated systems did work, on Vieques, buildings with old wiring blew out the brand-new Tesla batteries. Other solar installations became overgrown with weeds, which inhibited the effectiveness of the panels by blocking the sun. One solar installation was put on a hospital that never reopened. There was no maintenance. If something broke or did not work on the solar systems, it just stayed that way.[16] In a rush to make Vieques and Culebra "test cases," why did no one teach local residents to understand how to diagnose issues and maintain the equipment?

This type of a result is energy coloniality in action, where the actions taken might seem to be philanthropic and beneficial. However, the end result was extremely mixed, providing less benefit to local residents than a community-designed plan would have. For a lengthier discussion of training programs, see chapter 7 on Isabela, where another group brought solar equipment after the hurricane and also built a training program for island residents to become certified solar installers.

Community Solar

Community solar, in its broadest definition, means a system allowing greater access to and greater benefits from solar or other types of renewable energy. The Solar Energy Industry Association defines community solar as "equal access to the economic and environmental benefits of solar energy generation regardless of the physical attributes or ownership of their home or business."[17] As mentioned above, there are many definitions of community solar, and the precise way in which it is implemented depends upon a variety of factors. There are several working models which address a variety of limitations that restrict access to solar, ranging from the physical attributes of the building to cost. The building itself is a particularly relevant part of access.

Many people do not own their own roofs, such as renters or people who live in co-op buildings and apartments. Others live in buildings without enough sun to make rooftop solar viable or have other geographical or structural reasons because of which rooftop systems do not work. Community solar can address this by siting the solar panels on another building.

Another critical aspect of access is cost. Between purchasing and installing the equipment and providing the necessary maintenance, community solar provides greater affordability through group purchases, municipal, public, and/or non-profit ownership, financing at very low or zero interest rates, and other economic arrangements that extend the benefits of clean energy to people for whom it would otherwise be out of reach.

Finally, there are two infrequently mentioned but important advantages of community solar. They are:

1) The trust and confidence it builds,
2) The opportunity for education about energy.

In order to buy into any system, people need to trust that the equipment will work as intended, that the financing is fair and accurate, and that there will be no unpleasant surprises down the road. In order to continue these systems long term, they need to understand them and should be able to see this new system in their own future. Below, I describe the additional benefits of education and youth engagement that a community-oriented approach like that of Coquí Solar are in the process of forging.

Throughout Puerto Rico, after Hurricane Maria, many homes were damaged or destroyed, and rooftop solar was not viable simply because many families still did not have roofs or there was structural damage done to the home. Even 6 months to a year after the hurricane hit, blue tarps covered many homes, showing how extensive roof damage and also how insufficient support systems like insurance and FEMA are.[18] If residents of a low-income community like Salinas have to first repair their roof and the rest of their house, and only then can they install solar, the economic on-ramp to reliable and affordable electricity is significantly increased. Community solar located on other buildings is one way to allow those families to gain access to reliable electricity, which then becomes a means to support them as they work through all the necessary tasks to repair and restore home, work, and family life.

Ruth describes how an even larger community orientation took place to help restore these homes even prior to installing solar:

> *There's a lot of repairing homes, especially rooftops and homes that were damaged by the hurricane. There's a project called "Construyendo Solidaridad*

desde el Amor y la Entrega," which translates to "Building Solidarity through Love and Commitment." It is a mutual aid program and has to do with Coqui Solar because rooftops have to be in good condition to be able to do the installations. IDEBAJO has some funding to buy the construction materials, and the community helps to get the best prices . . . building materials are going up quite a bit after the hurricane. So people are getting together to buy the construction materials and then help each other out to rebuild their homes.[19]

In practical terms, community solar means that an organization, business, or other types of entities can build a large solar array situated where conditions are ideal: a large rooftop such as a warehouse, house of worship, school, or university. It could be installed on a series of small rooftops, such as a residential community, or elevated over a parking lot, which would also provide shade. Because Puerto Rico is a fairly small island, land use is an important concern, and Coqui Solar and Queremos Sol's proposals suggest preserving arable land and habitat and installing land-based solar installations on landfills or contaminated sites such as brownfields.

The governing organization for community solar can be a non-profit founded to manage the solar array or can be an existing entity that chooses to invest, for example, a government or community center. Typically, that organization then builds and owns the solar array, although sometimes the actual array is owned and maintained by the solar installation company. The space that the solar array is located on can either be rented or owned by the organization. The organization becomes a solar power developer by investing in the equipment and installation, and they make arrangements to ensure any needed maintenance.

In economic terms, community members can buy in and purchase shares of the electric power produced by the cooperatively-owned solar array. As non-profit entities, the advantage of community solar is to use an economy of scale to allow them to sell their members' electricity at a rate lower than the going utility rate, similar to a food cooperative having lower prices than mainstream grocery stores through bulk purchases. Community solar projects can be off-grid or attached to the grid. For off-grid community solar projects, a utility scale battery is required, which adds considerably to the cost. However, an off-grid system has the unique advantage of creating a power supply that is independent of a larger grid and thus not vulnerable to storm damage or issues with lines in other parts of the grid because off-grid systems can island off from the main power grid. With this kind of proper sizing of an off-grid system, based on anticipated use, it is quite possible to create a system that provides all of the electricity people need. (For more on community solar projects, including those in an urban context, see chapter 4, about community solar in New York, and chapter 8, about off-grid infrastructure in La Riviera, and chapter 9, about Highland Park, Michigan.)

Clean Power: Design and Advantages

For El Coquí Solar, rooftops throughout the neighborhood could house solar panels, which could be purchased at lower rates through bulk purchasing and installation, and financed through low-interest credit union loans. Once the equipment is paid for, the primary costs are delivering the power to the homes and system maintenance. If delivered through the utility grid, the cost of powering your home can be as low as the utilities' minimal hookup fee. If the power is delivered through a localized "micro-grid," there could be no cost. The sun charges the battery all day, and the homes can use the power into the night. Part of the Coquí Solar project was to understand proper sizing and build community knowledge of energy use:

> *We want to make sure that people understand . . . what we call energy literacy. It's evaluating what your usage is; we went through that stage, doing energy inventories, determining how you can conserve more energy, right? And how to be more efficient in your energy use, what is your minimal energy need, and what is your ideal sort of energy budget. We do all those steps before. We work with the University of Puerto Rico, Mayagüez, Professor Efraín O'Neill, and the Technological Institute in Guayama. We have a lot of alliances with different groups and organizations for this project.*[20]

In rural areas with a lot of sun, frequent power outages, and relatively modest energy needs, a community-owned solar system, particularly with its own microgrid and battery backup, makes a lot of sense. In areas like Salinas, six months after the second-biggest power outage the world has ever seen, with the perpetual delays in rebuilding the same grid, with a dubious design to begin with, and where electricity rates were the second-highest in the nation prior to the hurricane—this kind of energy system not only makes sense, but also is essential for survival.

Ruth described how they turned toward community solar:

> *We've looked at other places where communities have come together and purchased in bulk, you know, the PV equipment, and they get better prices. The small amount of renewable energy that we have here (on the island) has not been a very good deal for the Puerto Rican ratepayers. The contracts were not negotiated, with any transparency . . . no competitive bidding. Basically, contracts were given out at very high prices. . . . The companies that have done the rooftop solar here have imposed onerous terms on people in sort of constituting second mortgages on people's properties, a very high rate and not necessarily with a battery system to back it up. So during the hurricane, some people found that—although they had solar PV equipment on their rooftops—they were*

interconnected and didn't have any storage capabilities, and so they didn't have the power either, even with the panels.[21]

She expanded on the additional benefits of community-based systems:

We believe that the kinds of micro-grids that are best suited to Puerto Rico would be rooftop solar communities. If you're talking about energy generation closer to communities, it needs to be cleaner because you can't very well have a microgrid that burns fossil fuels, for example, close to a community. You're going to aggravate respiratory diseases and other kinds of ailments related to the fumes and the emissions from dirty power generation. With rooftop solar, for example, you don't have the need for as much of a distribution system, certainly no transmission over long distances. You don't need that if you're providing within a given community. So it really cuts down on transmission losses and expenses and also that to some extent on distribution.[22]

Given the question that this chapter began with—about the difference that rooftop solar and distributed generation could have made and how renewable energy sources could also be a solution to designing electrical systems to withstand storms—this shows a critical difference in how a power system centered on rooftop solar would improve safety and the restoration of power after hurricanes. Ruth continued:

If you look at the solar array that we have right next to the El Coquí community, it is totally unscathed, right? Whereas the power lines between Salinas and Guayama were totally torn down. I think that's what we saw generally, with the exception of the area where Hurricane Maria made landfall, in Yabucoa, and on the eastern coast in Humacao, where solar arrays and windmills there were destroyed. But that's because the hurricane made landfall there with very strong winds. But throughout the island, as you go westward, you see that the poles and the transmission towers were knocked down, but the solar equipment held up much better. (Santiago, 2018)

How Much Solar Power Does Puerto Rico Need?

One of the most common myths about solar is that it cannot provide enough energy to run a modern home or business. This could not be further from the truth, particularly on a Caribbean island with year-round, highly predictable sun. A 2009 study by researchers at the University of Puerto Rico Mayagüez, Agustin Irizarry-Rivera, Efrain O'Neill-Carrillo, and José Colucci-Rios, powering the island with solar is not only within reach, but it would also only require about two-thirds of the rooftops:

The ARET study, which are achievable renewable energy target study . . . indicating that photovoltaic (PV) installations on about 65% of the rooftops in Puerto Rico would provide well over a hundred percent of the energy needed on the whole island.[23]

This holds true in 2018; in fact, with panels becoming more efficient, Lionel Orama-Exclusa, energy expert at the University of Puerto Rico at Mayagüez, stated: "We have enough renewable resources—wind, solar, water [hydropower], and biomass—to energize twice our actual consumption. It's not that we can go 100 percent, we can even go 200 percent."[24]

Expanding on the idea of how much electricity island residents use, particularly in the year after the hurricane, with businesses, schools, government offices, and everything else which were closed for months, Ruth continued:

Keep in mind that in 2009 (when the ARET study was done), Puerto Rico's economy demanded much more energy generation than it does now. Right now, we have barely a peak demand of barely 3000 megawatts, probably a lot less after the hurricane. And our so-called "installed capacity" generation capacity is 5,836 megawatts. We have almost double the generation of the peak demand.[25]

This is an important consideration. Energy-hungry cities, particularly those in northern climates, would likely need to dedicate far more than 65 percent of their roofs to create all the electricity they use. But not so for Puerto Rico. In a 2019 Territory Profile and Energy Estimate whitepaper, the U.S. Energy Information Administration reported, "Puerto Rico's energy consumption per capita in recent years has been about one-third of the states' level."[26] This supports the idea that rooftop solar is not only a stronger system for withstanding hurricanes but that it is a viable means to produce all of the electricity that Puerto Rico's citizens need.

Notably, this same Energy Estimate whitepaper also states that, while power use is much less, "consumers in Puerto Rico pay one of the highest U.S. electricity rates." The viability of rooftop community solar in technical and pragmatic ways, combined with the existing high costs, is a deeply compelling argument. But another important reason compels the residents of Salinas and Guayama: living in the shadow of dirty power.

DIRTY POWER: APPLIED ENERGY SYSTEMS AND TOXIC WASTE

This crystal clear vision for Puerto Rico's future shows the many benefits that solar has, which go well beyond just providing electricity. In Salinas, the advocacy for solar *after* the hurricane is thrown into an entirely different context by their advocacy for solar *prior* to the hurricane. Both IDEBAJO and Dialogo conducted campaigns against toxic waste and pollution created from the towering power plants, which have caused elevated levels of cancer and respiratory disease in the area. Despite legal challenges, advocacy campaigns, and earlier tests showing the aquifer was polluted, a 2019 study showed contamination in the groundwater to be irreparable.[27] Coquí Solar was working from a vision of the clear advantages—and basing that vision in the lived experience of the dangers and consequences of fossil fuel power production to human health and to local habitats.

The southern coast of Puerto Rico is home to the Jobos Bay and the protected areas of the Jobos Bay National Estuarine Research Reserve. A protected estuary where research is done on endangered species, including the hawksbill turtle and the West Indian manatee,[28] Jobos Bay Estuary is also vital for a healthy local commercial and recreational fishing industry, marine recreation, and ecotourism.[29] Salinas is on the west side of the Jobos Bay, and Guayama is on the east. Home to Puerto Rico's two most contaminating power plants, with one becoming particularly notorious for turning its toxic waste into bricks for use in construction projects: the Applied Energy Systems (AES) Power Plant.[30]

Salinas, Guayama, and surrounding towns built community environmental initiatives for years. IDEBAJO was formally founded in 2010 to promote eco-development and community empowerment, although there were many previous initiatives, mostly organized by Nelson Santos and other local activists dating back to the early 1970s (Santiago, 2020).

There is a lot to defend: the rich fishing grounds of the Jobos Bay supported a local industry and provided a reliable food source in a region with few economic opportunities. With a wealth of ocean life in the Jobos Bay, Salinas and Guayama also had a stark daily reminder of environmental danger—the Bay is flanked by two power plants: in addition to the AES Power Plant in Guayama, which burns coal, the oil-burning Aguirre Power Station lies across the Bay. Everyone in the Salinas-Guayama region lives in the shadow of these plants.

The AES Power Plant has been involved in litigation for years about the disposal of its toxic coal ash. The utility had turned toxic coal ash into an alleged construction aggregate material which they called Agremax. In order to get Agremax into the hands of contractors, they provided free transportation to construction sites, charging exceptionally low rates for it (Santiago, 2020). This introduced Agremax for construction purposes, despite its origins as coal ash and subsequent high concentrations of heavy metals and carcinogenic toxins. It was used to build roads and other types of infrastructure. When it rained, the toxins in the Agremax seeped into the groundwater.

In Ruth's words:

> *The AES plant is a 454 megawatt, coal-burning power plant that opened up operations in November 2002. They were supposed to be a "clean coal type" technology plants. They were supposed to be zero water discharge, and they were supposed to have a beneficial use for all of their coal ash, because they never created, they never established a designated place for disposal of their coal combustion residuals. Unfortunately, none of that turned out to be true.*[31]

The promises made by these power plants had been made frequently in other places, with similar results. "Clean coal" does not have a clear definition and is more of a political buzzword that has encompassed different technologies over time.[32] Zero water discharge is a type of power plant technology that does not release contaminated water, following regulations around coal ash which were designed to protect water quality, including preventing heavy metal.[33]

These types of regulations, however, do have an effect, if followed, with the evidence present in our atmosphere. For example, satellites can measure sulfur dioxide levels from coal-burning power plants in other parts of the United States. Scientists can use these images to see significant reductions in as little as five years. This decline in pollution is largely attributed to clean air regulations (NASA, 2011).[34]

AES, however, attempted to skirt these obligations entirely and were ultimately stopped through litigation. Ruth continued:

> *Pretty soon after they started, they started accumulating the coal ash on site at the plant. Later they started to use it as fill material at different construction sites, mostly within the Guayama region . . . AES generates about 300,000 tons of coal combustion residuals . . . what we commonly call coal ash, what they call Agremax. Basically, after they combine the two—the fly ash and the bottom ash—they combine that, they hydrate it, and supposedly they dry it, cut it up into pieces, and then call it Agremax. The so-called Agremax, or these residuals, are a concentration of heavy metals, toxic heavy metals, because it's what's left after burning the coal. And so it's very concentrated, dangerous substances like cadmium, hexavalent chromium, arsenic, boron, molybdenum, and selenium.*[35]

Coal ash contamination is not limited to Puerto Rico. In fact, an assessment of coal ash contamination in the U.S. and territories, maps the locations nationally. Even the utility industry's own data measures the rate of groundwater contamination at 91 percent and states that 95 percent of sites dump coal ash into ponds with no liners to protect the groundwater.[36] In other words, even the utilities know that only 5 percent to 10 percent of the time do they protect residents from the toxic waste they produce.

Numerous studies from around the world have documented the toxicity of coal ash, including causing cancer, cognitive deficits, heart and lung damage, kidney disease, reproductive issues, impaired bone growth, early death, and increased infant mortality. In areas of the Dominican Republic with significant coal ash contamination, women gave birth to babies without vital organs or limbs;[37,38] and[39]. Other studies have found coal ash waste to be more radioactive than nuclear waste per unit of power produced[40] In Puerto Rico, research found that virtually 100 percent of coal residuals were used in constructing homes, businesses, and roads, with much of this construction material used in flood-prone areas in the southeastern region of the island, creating a means for the toxic waste to leach into water (Santiago, 2012).[41]

With this background knowledge, the community fought back:

> *We had lots of heavy metals and less toxic constituents that are dangerous to both public health and also the natural environment. The AES power plant never had a dedicated disposal facility. So what they did was initially tried . . . was to get people in other countries, especially, to take the AES coal ash under the name of an aggregate building material.*[42] *And not many people went for that. Although unfortunately, some of the AES coal ash was taken to the Dominican Republic and caused a fiasco there. There were two lawsuits, both of which AES ended up settling and paying millions of dollars to settle for the damages caused by their coal ash. They didn't admit damages, but those were the claims, and both lawsuits were settled. We have the settlement agreement for the first case. AES and its parent company paid $6 million to the government of the Dominican Republic. Later on, more recently, in 2014, they settled the second claim also. . . . This was published in Bloomberg News and they also had to pay.*[43]

From the Bloomberg News report: "AES Corp. settled a lawsuit accusing the power-generating company of allowing one of its units to dump coal ash on beaches in the Dominican Republic, which allegedly caused a spate of birth defects in children."[44] A global company, AES and its subsidiaries have also been disciplined in other countries: in 2019, they were fined $5 million by Chile for hindering the adoption of renewable energy.[45] Being blocked from dumping this ash in other places, including Orlando, Florida,[46] the solution that AES found was to heap the coal ash in one spot in their compound, forming a giant gray mountain some five stories tall. This coal ash mountain has been the subject of multiple investigative journalist pieces and has become notorious throughout the island.[47,48,49] Not only is the coal ash toxic—it is also mobile. It drifts through the region, traveling on otherwise pristine ocean breezes into yards, school grounds, and central plazas.

Ruth explained just how dangerous it is:

The aggregate and so-called aggregate material that they call Agremax is really nothing but coal ash. And as soon as it is in contact with the elements, either rain or the wind, or gets loaded into a truck, it turns back into the dusty ashy material that generates a huge amount of fugitive dust.[50]

With no cover and no lining underneath, it is entirely certain that the toxic waste is leaching into the Jobos Bay and into the aquifer. Ruth described driving to it after the hurricane:

A few days after Hurricane Maria, I drove over here to AES to see what was happening. And what I saw was a lot of flooding in the area because the plant is sited basically in a wetland, right? So that's just south of the Aguirre State Forest and right on the outer Jobos Bay or what is also referred to as Las Mareas Bay. That's all a wetland. There was a huge amount of water in the area adjacent to the plant, within the plant, et cetera. The plant was not operating, and it was very dark, but still, the coal ash pile was there, and it looked wet, obviously. What we saw was erosion. It looked like streaks of water on the different sides of the mountains.[51]

The Coal Combustion Residuals Rule requires testing groundwater. AES had to get the aquifer tested, and Ruth had received the results right before the community solar meeting in March 2018. During the meeting, she shared the results.

This week, we learned that in addition to the toxic emissions that are coming from the smokestacks, we learned that they are contaminating the aquifer with the coal ash mountains that are located on the plant grounds. This is an enormous mountain of coal ash and is contaminating the land in many places, and also the aquifer. This was documented from a study that they had to submit to the EPA as a consequence of a regulation that we spent a lot of time working on.[52,53] *Which the administration now wants to reverse.*

While this is bad news, there are positive aspects to this, having to do with a modicum of transparency:

At a minimum, it has allowed us to get this information, this evidence, and all that much more of a reason to continue with the momentum on the Coquí Solar project, recognizing that this way of generating energy is simply not in the best interests of these communities and of the people of Puerto Rico in general.

At one point in the EQB report (the environmental quality board) indicates that it was 50 feet high, 200 feet long, and two hundred feet wide. Now, the pile that we saw is a few years old . . . they haven't been able to use it as fill material since at least 2012.[54]

She continues on to explain the systemic damage this coal ash pile causes:

When it rains, that material seeps into the ground or drains out into the bay. What we received about a week ago . . . was a copy of the groundwater monitoring report procured by AES, by a company that they hired to do groundwater monitoring, as required by the Coal Combustion Residuals Rule . . . it shows definite contamination.[55] *When you look at the sampling wells that AES had to set up north of the coal ash mountain and south, you can see huge differences and huge increases in the wells that are south of the coal ash mountain. It shows the effect that the coal ash seeping into the ground and into the groundwater has on the quality of the water.*[56]

The storage of coal ash is governed by many laws. As with many utilities throughout the continental United States, the enforcement of these laws is lax. When ash is exposed to the elements, the steady bay breezes transport it throughout neighboring communities, where it lodges in residents' lungs. Storms and even light rain carried toxic ash into the Jobos Bay. Many families rely on the Bay for food and for livelihood, and the region has a large commercial and recreational fishing industry.

This damage may seem irreversible, and this system is as immutable as the power system seems to be. But neither is true. Ultimately, local activists and legal advocates succeeded in banning Agremax for use in construction in Puerto Rico. And while the coal ash mountain remained for years, at the time of writing this book, AES is now removing it, subsequent to a new 2020 law far exceeding the previous coal ash protections.[57]

Gathered together at the El Coquí Community Center on this Saturday morning, this entire group discussing community solar has to be seen in the context of people who had been protecting their water and air quality for years and who had felt, firsthand, the consequences of fossil fuel production through damage to their bodies, in their water and food that they grow, and in the fish they catch. Growing up while seeing the stacks, the plumes of smoke, knowing that if there is a water shortage in your home, it might be because the water is being used by a power plant—all this leads to seeing the absolute advantages of solar power, years before a Category 4 hurricane showed the world how frail a power system could be. Salinas, Guayama, and all of southeastern Puerto Rico have held this waste in their hands for decades. Even knowing that solar will not solve every energy problem—it might bring with it a whole new, unique set of challenges—from the shadow of a coal ash mountain and living with a contaminated aquifer, toxic building materials, and a notoriously indifferent utility, whatever challenges that solar might bring would seem a welcome relief.

Emergency Power and Solar Kits

During the March 2018 Coquí Solar meeting, they were about to receive a small, emergency-sized solar power and battery system, a donation from one of the many hurricane relief groups. This small system would include a battery backup and would be installed on the El Coquí Community Center building. They were awaiting a final date for installation, expected to be sometime in the next month, and this long-awaited infrastructure upgrade would accomplish a great deal. It would ensure a reliable electricity supply in the event of another blackout when hurricane season began in June. It would provide a close-up view for all the students in the room, who could attend, watch, and participate in the installation. It would also provide a further base for community education—with small solar grids being installed all over the island, the learning curve was proceeding quickly, and the public was hungry to learn more. Perhaps most of all, it would provide a measure of security to the constant anxiety wrought by the trauma of Hurricane Maria.

The day's agenda began with reviewing the emergency solar power system the El Coquí Community Center would receive in the next month, which was a relatively short discussion. In stark contrast to the narrative surrounding Tesla bringing solar panels, including the "test case" language and the rushed approach, the majority of this discussion focused on ensuring that the systems would provide the most benefit possible to the community. For example, the systems would be made available to the most vulnerable: the bedridden, the elderly, those who needed to store medicine. Finally, there was a lengthy discussion of how to ensure that the wave of solar addressed the deep economic issues and lack of career opportunities that pushed people to leave the island.

From that meeting, Ruth recapped their progress to date:

It is interesting that we have spent so much time with professors, particularly Efrain and a few others, doing all of the planning and studying this center, the transformers, the poles, and the substation. Perhaps we could focus more on exploring the models under which the community can organize to see how to acquire photovoltaic (PV) equipment and how to organize around it. Also, how can we create social impact using these systems?[58]

Six months later, in October 2018, in a follow-up interview, Ruth shared how much of this planning was realized in just a few months. They had already made connections with the Technological Institute of Guayama, which has a solar energy training program. They were also planning what those models would be and how to plan with the local university lab and other resources. The planning included sizing the solar power systems appropriate for budgets but large enough to be effective:

> *It is a solar installation of PV rooftop, but it has fewer panels and just enough to power the absolute basic necessities that a person might have in their home. It requires an energy inventory, we call it, where the participants go to the home and figure out what the energy usage is. What are all the appliances, which are efficient and which are not, and look at those kinds of things, and determine, according to the person's budget, what kind of solar installation they can have.*[59]

This innovative strategy adapted a similar type of assessment, the energy audit, which is a staple of planning a renewable energy project and can be done by a certified professional or by an individual. The "energy inventory" is slightly different. In that the audit requires certified personnel such as electricians and engineers, while the inventory can be done by participants in a training program. Coqui Solar's training program served many purposes: it was a hands-on experience for students, performed a useful part of the assessment process, brought down the costs and time, and provided a way for people in training to gauge the development of their skills and knowledge. This is a prime example of how a community-centered approach yields more benefits.

Throughout the industry, stakeholders are looking at how to design a solar PV system to be "hurricane-resilient." Various models and specifications attempt to establish a set of best practices as well as look at common reasons systems fail during hurricanes (Stone and Burgess, 2018).[60] For Coquí Solar participants, this assessment of mechanical parameters is valuable for them to understand the day-to-day application on relevant regulations and technology, so they can participate in this growing industry and possibly even lead the development of hurricane-proof solar power systems.

Coquí Solar had gone beyond training, however, and was already designing a solar kit:

> *They are connected with an inverter to . . . a few solar panels on your rooftop. And, since we're talking about something that's going to be very accessible to most people, even very low-income people, we're talking about solar kits; that's our next step. Installing solar kits within the community and taking into account that some people have special medical needs, and there's a social purpose to*

this, right? People who have those needs are a priority. What we're thinking is that people can get small loans at the credit unions in town. They request whatever they think they can afford monthly, and on the basis of that, they get their installation. But there will be cases where people have special medical needs or special situations that require more, a bigger installation than they think they can afford, and that's when the social considerations come in, because we are getting funding to do some of this work.[61]

IDEBAJO and Coquí Solar had gained a reputation for their work, and after the wave of donated equipment came another wave of resources. Ruth continued:

The fundraising for the solar community was started, but it was very slow before the hurricane. Very, very slow. So much so that we were not able to get funding until after the hurricane to do the installation for the community center pilot project.

When asked if their plans helped attract funding for solar projects, given how clear the need for renewable energy became, Ruth answered:

Absolutely. Right away. It was amazing. We didn't even ask for it. Fundación Segarra Boerman came to us, said "we know that you're doing this, that you're planning this project," here's the funding.[62]

Their work attracted more than funding—Coquí Solar also began working with a professor and group of students from MIT.[63] This group came as part of a practicum course in the MIT Department of Urban Studies and Planning, bringing over a dozen students from Cambridge to Salinas in the fall of 2018. This collaboration allowed these burgeoning professionals to see how dirty power and clean power have come together in Salinas. They wrote, after the trip: "The environmental justice concerns are also very real and tangible: community members showed us pictures of coal ash in their gardens that had blown over from an improperly stored coal ash pile at one of the nearby power plants."[64]

With this influx of support, and with the shadow of AES and the Aguirre power plant still falling on El Coquí's neighborhood, the focus expanded to show how even more people could benefit. The term "prosumer" was introduced into expert witness testimony by Professor Agustín Irizarry-Rivera during a motion to submit testimony to the Puerto Rico Public Service Regulatory Board and the Puerto Rico Energy Bureau, in October 2019, regarding the PREPA Integrated Resource Plan:

I think we need to put all the pieces together so it's like a pipeline or something that's easily done . . . so that people can actually finance their systems. Because very few people here have enough money to do even a solar kit without some kind of loan or grant. Now that the technology is available, people need to have energy closer to where they can manage it and become prosumers.[65]

A prosumer is a customer with "the capacity to generate electric power for self-consumption that, in turn, have the capacity to supply any energy surplus through the electric power grid." In other words, people create their own power and sell it back to the grid—a relatively common practice in states with strong solar programs. Consider the distance traveled from "energy coloniality," in which the consumer has little to no agency to decide how to power their home, and thus loses power on a larger scale, to prosumer, where the individual retains a sense of agency, and also benefits from power production.

In the October follow-up interview, I asked Ruth about this larger scale change:

You can tell the people are much more energy literate. They're much more aware. I am so amazed. Just after the hurricane, people started moving around. There were fuel shortages, and the generators were not that all that helpful, and actually very toxic to some people. The noise was unbearable. So people started looking for options and getting into renewable energy, especially solar panels. It's a sea change. It's absolutely different. The Association of Solar Energy Consultants and Contractors here in Puerto Rico are telling us that they're seeing so many times over the number of installations that they're doing now that they compared to what they were doing before the hurricane.[66]

This knowledge building and sense of forward progress show how quickly results can materialize. Underlying this progress were the as-yet unsolved larger problems, as knowledge and an emergency solar array can significantly ease tension, however that does not change the underlying situation nor that the people "are paying the price for the corruption." She describes the lingering fears and trauma:

Actually, you know, we've had a couple of scares, and I think that's how it's going to be in the future. People are traumatized. I remember on April 18th [2018], when we had the three-day blackout . . . the transmission lines from AES went down, and it tripped the whole system out.[67] *People went into hysteria. People were at the gas stations and buying water . . . just in a generalized trauma.*[68]

QUEREMOS SOL—"WE WANT SOLAR"

In 2018, as Puerto Rico rebuilt its grid and the Puerto Rican Legislature debated the new energy law, Coquí Solar and other solar advocates on the island took their next step. A group of 16 individuals and ten organizations formed Queremos Sol ("We Want Solar") and drafted a detailed proposal of their energy vision and energy plan. This proposal was presented to the legislature in the months preceding a vote on a new energy policy.[69] Among the organizational members was the PREPA union, which is interesting that the utility workers also want to see a transformation in the energy system they built and maintain.

Queremos Sol's community-based proposal for solar was presented to the Puerto Rican legislature as they developed the new energy law and detailed the vastly larger benefits possible from solar power, as well as the design for an energy system that included microgrids, battery storage, distributed generation, and the role PREPA could play in an improved system. Presenting this proposal to the legislature was a major step for the island's solar advocates, establishing their collective voice and galvanizing what had been a widespread collection of grassroots organizations into a more unified network.

Around the same time, this coalition, along with many other solar advocates, planned their first island-wide gathering. As mentioned earlier in this article, the Transformación Energética Desde Las Comunidades, or Community Energy Transformation Gathering, took place at the Interamerican Law School in San Juan on December 7 and 8, 2018. Further galvanizing this collection of organizations, scholars, activists, lawyers, and other stakeholders, this event also had attendees from Colombia and the Dominican Republic, showing how broadly this change was desired through other Caribbean and Latin American stakeholders. Among the many presenters at this event were youth representatives from Coquí Solar, now with their own branded t-shirts, and presenting to international conference participants about their work.

NEW ENERGY LAW, NEW CHALLENGES

On April 11, 2019, the Puerto Rican legislature passed PS 1121, or Act 17,[70]a new energy bill with ambitious and laudable renewable energy goals, including 40 percent renewable energy by 2025 and 100 percent by 2050. The law also mandated eliminating coal by 2028. This bill was signed into a law later by then Governor Rosselló. It allowed microgrids, net metering and eliminated a lengthy permitting process for small installations such as solar kits and emergency or single-home-sized systems. This expedited the installation process considerably.

These are hard-fought and well-deserved technical and regulatory advancements. At the time of passing, Puerto Rico got 2 percent of its energy from renewables; this law mandated a change to 40 percent in the space of 5 years. While some press reports at the time reported that these goals had no clear path to achievement, the Queremos Sol detailed proposal presented clear options. The renewable energy community expressed excitement over this stated goal, although many were skeptical of the timeline and true commitment. Puerto Rico's solar advocates had come far and had succeeded in bringing a perspective that united community benefit, economic gain, and broader empowerment to the public, many of whom were witnessing firsthand the improvements in their energy system. Act 17 was a milestone and a gateway to the next chapter. Solar advocates were going to have to defend their vision and continue the hard-fought progress for solar.

The developments for Coquí Solar, and solar in general, continue. Even as the entire industry of solar experts, finance companies, developers, and installers lined up, ready to cover Puerto Rico's rooftops with solar arrays and realize the promises of their particular clean energy revolution. Solar has come to stay in Puerto Rico. The young people from El Coquí, and other areas, are vastly more competent and knowledgeable about solar. They have exponentially more access to it than their professors and advisors did at the same age. New leaders are developing their own voices and networks, like the members of Queremos Sol, Coquí Solar. And now, they have laws that open pathways to clean energy that previously did not exist.

Natural Gas

At the time of writing this, two significant challenges threatened to undermine the extensive and visionary work of the movement for solar. One is natural gas. When Act 17 was passed, the Legislature also included using natural methane gas as a transition fuel. As PREPA developed its plan following this legislation, natural gas continued to receive more attention than many felt was necessary. After this law was passed, Rocky Mountain Institute, an independent, nonpartisan non-profit focused on energy transformation, which was involved in the ongoing development of Puerto Rico's energy system, reported on the shortfalls in PREPA's plan, including a clear bias toward natural gas over renewables.[71]

Conflicts of Interest

Other emerging challenges like the potential for conflicts of interest, specifically in regards to Siemens, a German multinational conglomerate company and the largest industrial manufacturing company in Europe, and funding. Ruth summed these up as:

Siemens has a conflict of interest here because Siemens can very easily say, oh, you should have these many new combined-cycle units to facilitate the integration of renewable energy.[72, 73] *Because that's how it's being framed. It's like . . . you do natural gas as a transition, but we know that ultimately we're going to go to high levels of renewable energy. Government is going in the wrong direction here . . . In [Siemens'] presentation, one of the pages indicates that the government is proposing to have four or five new natural gas projects, including pipelines, offshore gas ports, and land-based gas plants, basically all over the island. In the meantime, where is the funding?*[74]

She noted the report by New York Governor Andrew Cuomo's office, "Build It Back Better,"[75] written in December 2017, which recommended extensive rebuilding plans, and sketched out $17 billion in costs. Of this $17 billion, only $1.4 billion—8 percent—was earmarked for renewables. These numbers raise an important question for ongoing research: How can Puerto Rico get to 40 percent of the electricity from renewables unless 40 percent of the funding is earmarked for renewables? An excellent area for ongoing research is to follow the growing conflict between use of renewable energy and natural gas, which was clear in the Integrated Resource Plan (IRP) presented by PREPA in 2020.[76] The Queremos Sol plan challenges this IRP with their plan explaining how the island can meet 100 percent of its energy needs with distributed solar and storage.

Solar Tax

With all this progress and promise, it is easy to see that a 100 percent clean energy future is quite possible for Puerto Rico. It is equally easy to see how difficult it might be to get there. More challenges continue to emerge. For example, in 2019, a debt-restructuring proposal included provisions hindering renewable energy by creating additional fees for people who are using solar panels. Widely criticized as a "solar tax," the growing number of solar stakeholders challenged the implementation of these fees. (Lloréns Velez, 2019 and SEIA Press Release, 2019)[77,78] At the time of writing, these fees were not passed. However, throughout the United States, similar fees and tax-like laws have been proposed, as well as utilities moving to reduce promised net metering rates, which undermine the economic incentives of solar. These are regularly proposed and equally regularly fought. Sometimes they are blocked, and sometimes they are passed by state legislatures.

Extenuating circumstances with natural gas production continue to threaten the future of clean energy on the island. In January 2020, a combination of a 6.4 magnitude earthquake and numerous aftershocks damaged a natural gas power plant in Costa Sur and left many without power. This was followed quickly by March's Covid-19 and the ensuing quarantine, which reduced energy demand by approximately 10 percent. Despite reduced demand, PREPA asked for $1.26 billion in federal emergency funds to cover energy generation to replace the energy lost by the Costa Sur plant going down. If PREPA had received this funding, a New York company was shortlisted for the contract. A community-centered proposal that was in line with the established goals for the island would prioritize energy funding to building renewable energy infrastructure and establishing contractual agreements, giving priority to Puerto Rican-based companies and creating jobs on the island.

Widely opposed, this proposal did not move forward.[79] Controversy continues to characterize the privatization of Puerto Rico's grid, along with ongoing contract agreements for building natural gas infrastructure, which also met with successful challenges to the multimillion-and billion-dollar contracts.[80] These many varied challenges signal the level of vigilance needed to ensure compliance with a promise of the energy law and the 100 percent renewable energy goal.

CONCLUSION

Puerto Ricans, such as the members of Queremos Sol and Coquí Solar, know how to transition their energy grid and energy governance. Their plans and proposals were developed and tested by leading authorities who live on the island and who are committed to improving the economy and opportunities, as well as energy systems. They were supported by allies from the mainland and other parts of the Caribbean.

If political leaders choose to follow where the people lead, based on years of research and with an overall goal to develop the human infrastructure in Puerto Rico along with the technical infrastructure, this small island could lead the world in clean energy development.

ACKNOWLEDGMENTS

Coquí Solar was founded with the efforts of many people, including, as mentioned above: the Coqui Community Board and IDEBAJO members including Nelson Santos, Ismenia Figueroa, Roberto Thomas and others, as well as Prof. Efrain O'Neill Carrillo, Prof. Marcel Castro, Prof. Marla Perez and Prof. Cecil Ortiz.

NOTES

1. The U.S. Energy Information Administration. (EIA). "Puerto Rico's Electricity Service is Slow to Return after Hurricane Maria." U.S. EIA. October 2017
2. IDEBAJO's work on solar pre-Hurricane Maria: https://www.mariafund.org/idebajo; and their history with ecotourism: https://www.sapiens.org/culture/jobos-bay-community-activist.
3. Earthjustice. "Ruth Santiago, Clean Air Ambassador Biography." *Earthjustice.org*. 2019. https://earthjustice.org/50states/2013/ruth-santiago.
4. ANDA Asociación Nacional de Derecho Ambiental, and Enlace Latino de Acción Climática. 2018. "Event Listing: Transformación Energética Desde Las Comunidades (Community-Based Energy Transformation)." December 7, 2018. https://www.facebook.com/events/286994762144014/.
5. Santiago, Ruth. "How to Power a City." Interview by Melanie La Rosa. March, 2018.
6. Kern, Rebecca. "Rooftop Solar Nearly Doubles in Puerto Rico One Year after Maria." *Bloomberglaw.com. Bloomberg Law*, September 20, 2018. https://news.bloomberglaw.com/environment-and-energy/rooftop-solar-nearly-doubles-in-puerto-rico-one-year-after-maria.
7. LaRosa field notes, 2018. https://www.yarotek.com/yarotek-commits-to-local-communities-after-irma-and-maria
8. Santiago, Ruth. "How to Power a City." Interview by Melanie La Rosa. March, 2018.
9. Klein, Naomi. *The Shock Doctrine: The Rise of Disaster Capitalism*. Picador. 2007.
10. de Onís, Catalina M. "Energy Colonialism Powers the Ongoing Unnatural Disaster in Puerto Rico." *Frontiers in Communication* (2018). https://www.frontiersin.org/articles/10.3389/fcomm.2018.00002/full
11. Hinojosa, Jennifer, Edwin Meléndez, and K. Severino Pietri. "Population Decline and School Closure in Puerto Rico." *Center for Puerto Rican Studies at Hunter College* (2019). https://centropr.hunter.cuny.edu/sites/default/files/PDF_Publications/centro_rb2019-01_cor.pdf
12. Katz, Jonathan. "The Disappearing Schools of Puerto Rico." *New York Times*, Sept. 12, 2019. The Disappearing Schools of Puerto Rico.
13. U.S. Census. "The Puerto Rico Community Survey Annual Data" and "More Puerto Ricans Move to Mainland United States, Poverty Declines." U.S. Census Bureau. September 26, 2019.
14. Flores, Antonio, and Krogstad, Jens Manuel. "Puerto Rico's Population Declined Sharply after Hurricanes Maria and Irma." *Pew Research Center*. July 26, 2019.
15. DeWaard, J., J.E. Johnson, and Whitaker, S.D. "Out-migration from and Return Migration to Puerto Rico after Hurricane Maria: Evidence from the Consumer Credit Panel." *Population and Environment* 42 (2020), 28–42. doi: 10.1007/s11111-020-00339-5.

16. Kaufman, Alexander. "On Puerto Rico's 'Forgotten Island,' Tesla's Busted Solar Panels Tell A Cautionary Tale." *Huffpost.Com*, May 17, 2019. https://www.huffpost.com/entry/elon-musk-tesla-puerto-rico-renewable-energy_n_5ca51e99e4b082d775dfec35.

17. SEIA. "Community Solar." *Solar Energy Industries Association* webpage, accessed June 22, 2020. https://www.seia.org/initiatives/community-solar

18. Hernández, Arelis R. 2020. "Puerto Ricans Still Waiting on Disaster Funds as Hurricane Maria's Aftermath, Earthquakes Continue to Affect Life on the Island." *Washington Post*, January 9, 2020. https://www.washingtonpost.com/national/puerto-ricans-still-waiting-on-disaster-funds-as-hurricane-marias-aftermath-earthquakes-continue-to-affect-life-on-the-island/2020/01/19/3864fcea-387f-11ea-bb7b-265f4554af6d_story.html.

19. Santiago, Ruth. "How to Power a City." Interview by Melanie La Rosa. December, 2018.

20. Santiago, Ruth. "How to Power a City." Interview by Melanie La Rosa. December, 2018.

21. *EIA Electricity Sales Data for Puerto Rico Show Rate of Recovery Since Hurricanes*, https://www.eia.gov/todayinenergy/detail.php?id=36832#:~:text=Puerto%20Rico%20residential%20electricity%20customers,of%2013%20cents%20per%20kWh.

22. Santiago, Ruth. "How to Power a City." Interview by Melanie La Rosa. December, 2018.

23. Irizarry, A., B. Colucci, and Efrain O'Neill. "Achievable Renewable Energy Targets for Puerto Rico's Renewable Energy Portfolio Standard." *Puerto Rico Energy Affairs* (2008). https://bibliotecalegalambiental.files.wordpress.com/2013/12/achievable-renewableenergy-targets-fo-p-r.pdf and https://www.uprm.edu/aret/

24. Bagley, Katherine. "After the Storm, Puerto Rico Misses Chance to Rebuild with Renewables." *Yale E360*, May 31, 2018. https://e360.yale.edu/features/after-the-storm-puerto-rico-misses-a-chance-to-rebuild-with-renewables-hurricane-maria

25. Santiago, Ruth. "How to Power a City." Interview by Melanie La Rosa. March, 2018.

26. U.S. Energy Information Administration. "Puerto Rico Profile, Territory Profile and Energy Estimates." Accessed July 2, 2020. https://www.eia.gov/state/analysis.php?sid=RQ

27. Alfonso, Omar. "Arroyo Barril: Coal Ash And Death Remain 15 Years Later." *Periodismo Investigativo,* December 20, 2018.Azzopardi, Tom. "AES Corp. Fined $5 Million in Chile Over Coal Plant Data." *Bloomberg Law*, June 27, 2019. https://news.bloomberglaw.com/environment-and-energy/aes-corp-fined-5-million-in-chile-over-coal-misstatement

28. Field, Ralph, Eddie Laboy, Jorge Capellla, Pedro Robles, Carmen González, and Angel Dieppa. 2008. "Jobos Bay Estuarine Profile: A National Estuarine Research Reserve." *Coast.Noaa.gov*. U.S. National Oceanic and Atmospheric Administration. https://coast.noaa.gov/data/docs/nerrs/Reserves_JOB_SiteProfile.pdf.

29. NOAA Office for Coastal Management. "Jobos Bay National Estuarine Research Reserve." *National Oceanic and Atmospheric Administration* website. https://coast.noaa.gov/data/docs/nerrs/Handout-Jobos-Bay.pdf.

30. Alfonso, Omar "Damage by Coal Ash to the Southern Aquifer Cannot be Undone." *Center for Investigative Journalism.* March 25, 2019.

31. Santiago, Ruth. "How to Power a City." Interview by Melanie La Rosa. March, 2018.

32. Grossman, David. "Clean Coal Explained - What Is Clean Coal?" *Popular Mechanics*, November 13, 2020.

33. WesTech Engineering. "Wastewater Treatment for Power Plants: Considering Zero Liquid Discharge," www.WesTech-inc.com. September 5, 2017.

34. NASA. "NASA Satellite Confirms Sharp Decline in Pollution from US Coal Power Plants," NASA Goddard Science Center. December 01, 2011.

35. Santiago, Ruth. "How to Power a City." Interview by Melanie La Rosa. March, 2018.

36. Earthjustice. "Mapping the Coal Ash Contamination." *Earthjustice.org*. October 6, 2020. https://earthjustice.org/features/map-coal-ash-contaminated-sites

37. Hashan, Mahamudul, M. Farhad Howladar, Labiba Nusrat Jahan, and Pulok Kanti Deb. "Ash Content and its Relevance with the Coal Grade and Environment in Bangladesh." *International Journal of Science and Engineering Research* 4, no. 4 (2013), 669–676. https://www.researchgate.net/profile/Pulok_Deb/publication/236590642_Ash_Content_and_Its_Relevance_with_the_Coal_Grade_and_Environment_in_Bangladesh/links/02e7e51821c754d723000000/Ash-Content-and-Its-Relevance-with-the-Coal-Grade-and-Environment-in-Bangladesh.pdf

38. de Onis, Kathleen. *Spreading Toxicity: Illegal Coal Ash Disposal Practices in the Caribbean.* Ebook (2015). Indiana University. https://theieca.org/sites/default/files/conference-presentations/coce_2015_boulder/de_onis_kathleen_-612255293.pdf.

39. Kravchenko, Julia, and H. Kim Lyerly. "The Impact of Coal-powered Electrical Plants and Coal Ash Impoundments on the Health of Residential Communities." *North Carolina Medical Journal* 79, no. 5 (2018), 289–300. https://www.ncmedical-journal.com/content/79/5/289

40. Hvistendahl, Mara. "Coal Ash Is More Radioactive Than Nuclear Waste." *Scientific American*, December 13, 2007.

41. Santiago, Ruth. "Imminent and Substantial Endangerment to Human Health and the Environment from Use of Coal Ash as Fill Material at Construction Sites in Puerto Rico: A Case Study." *Procedia-Social and Behavioral Sciences* 37 (2012), 389–396. https://www.researchgate.net/publication/257715914_Imminent_and_Substantial_Endangerment_to_Human_Health_and_The_Environment_from_Use_of_Coal_Ash_as_fill_Material_at_Construction_Sites_in_Puerto_Rico_A_Case_Study

42. Santiago, Ruth, 2012.

43. Santiago, Ruth. "How to Power a City." Interview by Melanie La Rosa. March, 2018.

44. Feeley, Jef, and Chediak, Mark. "Power Company AES Settles Claims That It Killed or Deformed Babies With Dumped Coal Ash." *Bloomberg News*. April

4, 2016, https://www.bloomberg.com/news/articles/2016-04-04/aes-settles-suit-over-coal-ash-dumping-in-dominican-republic

45. Azzopardi, Tom. "AES Corp. Fined $5 Million in Chile Over Coal Plant Data." *Bloomberg Law*, June 27, 2019. https://news.bloomberglaw.com/environment-and-energy/aes-corp-fined-5-million-in-chile-over-coal-misstatement

46. Spear, Kevin. "Osceola County Landfill Takes In Coal Ash From Puerto Rico, Triggering Public Backlash." *Orlando Sentinel*, May 13, 2019.

47. Alfonso, Omar. "Arroyo Barril: Coal Ash And Death Remain 15 Years Later." *Periodismo Investigativo,* December 20, 2018. https://www.pbs.org/newshour/health/residents-of-this-city-already-worried-about-the-coal-burning-plant-nearby-then-came-hurricane-maria

48. Kelkar, Kamala, Ivette Feliciano, Zachary Green. "Residents of This City Already Worried About the Coal-burning Plant Nearby. Then Came Hurricane Maria." *PBS Newshour*, April 28, 2018. https://www.pbs.org/newshour/health/residents-of-this-city-already-worried-about-the-coal-burning-plant-nearby-then-came-hurricane-maria

49. Arroyo Barril: coal ash and death remain 15 years later

50. Santiago, 2018.

51. Santiago, 2018.

52. Garrabrants, A. C., D. S. Kosson, R. DeLapp, Peter Kariher, and Susan A. Thorneloe. "Leaching Behavior of "AGREMAX" Collected from a Coal-Fired Power Plant in Puerto Rico." 2012. https://cfpub.epa.gov/si/si_public_record_report.cfm?Lab=NRMRL&dirEntryId=307594

53. Alfonso, Omar. "EPA Adopts New Rules Tailored for AES Coal Plant Puerto Rico." *La Perla del Sur y Centro de Periodismo Investigativo.* August 3, 2018.

54. Garrabrants, et al. 2012.

55. U.S. EPA. "Disposal of Coal Combustion Residuals from Electric Utilities Rulemakings." EPA website. https://www.epa.gov/coalash/coal-ash-rule

56. Santiago, 2018.

57. Surrusco, Emilie Karrick. "In the Fight to Clean up Coal Ash, These States Are Making Progress." *EarthJustice*, February 6, 2020. https://earthjustice.org/blog/2020-february/in-the-fight-to-clean-up-coal-ash-these-states-are-making-progress

58. Santiago, March 2018 interview

59. Santiago, March 2018 interview

60. Stone, Laurie, and Burgess, Christopher. "Solar Under Storm: Designing Hurricane-Resilient PV Systems." *Rocky Mountain Institute* website. June 20, 2018. https://rmi.org/solar-under-storm-designing-hurricane-resilient-pv-systems/

61. Santiago, 2018

62. La Fundación Segarra Boerman e Hijos, Inc. "Fundación Segarra Boerman E Hijos, Inc." *Fsbpr.org. Fundación Segarra Boerman e Hijos, Inc.* Accessed June 17, 2020. https://www.fsbpr.org.

63. Santiago, 2018

64. Swain, Marian, Hsu, David, and Bui, Lily. "Developing a Resilient Energy Infrastructure for Puerto Rico." *MIT, Department of Urban Planning* website. 2018.

https://dusp.mit.edu/epp/news/developing-resilient-energy-infrastructure-puerto-rico

65. Puerto Rico Energy Bureau. "Testimony of Professor Agustín Irizarry-Rivera,Page 9, definition of "prosumer." *Energia.pr.gov.* 2019. https://energia.pr.gov/wp-content/uploads/2019/10/LEOs-Motion-for-Submission-of-Testimony-with-Testimonies.pdf

66. Santiago, 2018

67. Wagner, James and Robles, Frances. "Puerto Rico Is Once Again Hit by an Islandwide Blackout." *New York Times.* April 18, 2018. https://www.nytimes.com/2018/04/18/us/puerto-rico-power-outage.html

68. Santiago, 2018

69. Queremos Sol (We Want Solar). "Queremos Sol Propuesta, Versión 4.0 (We Want Solar Proposal v. 4.0)." *Queremossol.* February 1, 2020. https://www.queremossolpr.com/.

70. Legislature of Puerto Rico. "Act No. 17-2019." *Autoridad de Energia Eléctrica, Aeepr.com.* April 11, 2019.

71. Torbert, Roy, and Mike Henchen. "Implementing Puerto Rico's Energy Transformation." *Rocky Mountain Institute*, March 10, 2020. https://rmi.org/implementing-puerto-ricos-energy-transformation/

72. Siemens Energy. "Combined Heat and Power Brochure." *Siemens-Energy.com* (Global Website). Siemens. n.d. Accessed January 22, 2020. https://www.siemens-energy.com/global/en/offerings/power-generation/power-plants/combined-heat-and-power.html.

73. William, Driscoll. 2018. "Puerto Rico Utility Favors LNG over Solar in Siemens Plan." *PV Magazine*, November 14, 2018. https://www.pv-magazine.com/2018/11/14/puerto-rico-utility-favors-lng-over-solar-in-siemens-plan/.

74. Santiago, 2018

75. New York Power Authority, Puerto Rico Electric Power Authority, Puerto Rico Energy Commission, Consolidated Edison, Edison International, Electric Power Research Institute, Long Island Power Authority, and et al. 2017. "Build Back Better: Reimagining and Strengthening the Power Grid of Puerto Rico." *Governor.ny.gov.* New York State. https://www.governor.ny.gov/sites/default/files/atoms/files/PRERWG_Report_PR_Grid_Resiliency_Report.pdf#:~:text=On%20behalf%20of%20the%20Working.

76. Earthjustice. "Groups Argue for 100% Renewable Energy to the Puerto Rico Energy Bureau." *Earthjustice.org.* February 10, 2020. https://earthjustice.org/news/press/2020/100-percent-renewable-energy-governing-board-of-the-puerto-rico-electric-power-authority

77. Lloréns-Vélez, Eva. "Puerto Rico senator rails against utility's restructuring support agreement." *Caribbean Business.* May 14, 2019. https://caribbeanbusiness.com/puerto-rico-senator-rails-against-utilitys-restructuring-support-agreement/?cn-reloaded=1

78. SEIA. "Solar Industry Urges U.S. District Court to Reject Discriminatory Charges on Puerto Rico Solar Customers." Press Release, *Solar Energy Industries*

Association. October 25, 2019. https://www.seia.org/news/solar-industry-urges-us-district-court-reject-discriminatory-charges-puerto-rico-solar

79. Santiago, Ruth, de Onís, Catalina M., Cataldo, Kenji, and Lloréns, Hilda. "A Disastrous Methane Gas Scheme Threatens Puerto Rico's Energy Future." *Nacla.org*., June 4, 2020. https://nacla.org/news/2020/06/04/methane-gas-scheme-puerto-rico-energy

80. Gallucci, Maria. "The Privatization of Puerto Rico's Power Grid Is Mired in Controversy." *IEEE Spectrum*. July 8, 2020. https://spectrum.ieee.org/energywise/energy/policy/the-privatization-of-puerto-rico-power-grid-mired-in-controversy

Chapter 7

Isabela, Puerto Rico
Design for Survival

INTRODUCTION

As hurricanes and wildfires batter cities throughout the United States and the world, we repeatedly see one massive power grid failure after another. As a result, the vulnerability of our electric grid is evident, and its flaws are a common headline topic, from lines collapsing to transformer explosions, the grid plays a role in starting wildfires, and in rolling blackouts.[1]

"Grid failure" essentially means the power lines fell while the power plant creating the electricity is still operating. It has become an almost expected outcome of hurricanes, blizzards, and even thunderstorms. Response to grid failure typically includes utilities rebuilding the lines. Depending on the extent and cause of the grid failure, they might even "strengthen and harden" the power lines, essentially fixing the grid with upgraded technology to make them strong enough to withstand storms. Sometimes this works. Sometimes it does not, and a newly "hardened" system might fail again. Even the utility industry acknowledges an overwhelming need to harden the grid, particularly in light of the many recent massive grid failures in Puerto Rico, Texas, New York, and other states.[2] Typically, the utility industry passes the cost of building new lines on to the consumers, which adds to the cost of electricity. Many in the utility field suggest that the expense of strengthening and hardening the lines should not fall on the consumers.[3] (See chapter 5 on Vermont for a longer discussion of strengthening and hardening the grid.)

What if there was another way? What if, instead of rebuilding something that falls down regularly, we redesigned it? What if each time a grid falls, its replacement was a new system designed and built better so that the system, and the people it supports, can survive storms?

This chapter looks at a different response to grid failure and loss of power, which is the transformation of the design of home power systems by installing a small solar-and-battery power system that will withstand storms. Although small, these systems can power homes at normal levels, will cost nothing once they are installed, and will function immediately after a storm is over.

The actions portrayed in this chapter began after two of the nation's most destructive and dangerous grid failures: after 2012's Hurricane Sandy, which knocked out power from Long Island, through parts of New York City, and down the New Jersey coast; and after 2017's Hurricane Maria, which slammed into Puerto Rico with winds over 100 miles per hour, causing an entire island of 3.4 million people to lose power for a year, the second-longest power outage in world history.

In areas with many hurricanes, grid failure and power outages are frequent. By improving energy systems' design and architecture, such as building microgrids and off-grid solar power systems with backup batteries, we can build homes, businesses, and communities that survive hurricanes. In addition, we know that electrical power production from fossil fuels is one of the most significant causes of climate change, making hurricanes and heat waves worse, leading to power outages and the destruction of homes and communities. Creating solar-powered buildings that can withstand storms and which do not contribute to pollution and climate change is a way to stop this vicious cycle. They will also significantly improve safety, save money, reduce and even eliminate fossil fuel use.

CHAPTER SUMMARY

This chapter follows a family based in Brooklyn and Isabela, Puerto Rico, as they create a humanitarian group to build emergency solar power systems in Puerto Rico. Then, they founded an industry-standard solar training program for island residents. They based this on their previous work building emergency sized solar power generators in New York after Hurricane Sandy and applied the same successful design to Puerto Rico after Hurricane Maria.

On a technical level, this chapter looks at the role of solar power and its role in an improved design for household power systems that can operate despite, during, and immediately after storms. It explores how this type of rooftop-based design minimizes power lines in hurricane-prone areas, instead using community-based renewable power to create more robust systems. Finally, it looks at human systems as well as technical ones: how can training programs build the capacity of local communities to install, manage, and maintain their own solar power systems? Is installing equipment enough? Or, are broader education and training programs necessary for the total transformation of power systems?

Overall, this chapter addresses several questions: how can the design of home power systems evolve to be safer and more resilient? What are the advantages of smaller systems over large ones? Finally, how can energy system transformation strengthen community control over those systems, ensure that people save money, and create long-term careers?

METHODOLOGY

As mentioned in chapter 6 on Salinas, when I heard the news about Hurricane Maria's devastation to Puerto Rico, I started looking for people turning to solar as a solution. It was not a question of *whether* anyone was doing this; it was a question of *who* and *how* it was taking place.

I scoured the news and came across a Rockaways-based media outlet reporting on a Brooklyn-based urban design and architecture firm building solar power generators to bring to Puerto Rico. I found the firm's website—Local Office Landscape Architecture—and contacted Jennifer Bolstad and Walter Meyer, who responded immediately. They shared their planned solar relief efforts, which were currently being organized at their Brooklyn office. Ultimately, this group would become ¡Solar Libre! Puerto Rico.

Filming began on a weekend early in October 2017, two weeks after Hurricane Maria, with volunteers sorting, testing, and loading donated solar panels for shipping to Puerto Rico. They were planning a series of relief trips to Puerto Rico, with plans to leave as soon as possible, bringing the gear, assessing sites to donate the solar-and-battery systems. Then, they would set up their headquarters in a bed and breakfast owned by Thomas Meyer, Walter's father.

A former student of mine, now a cinematographer, volunteered to accompany them on the first relief trip, leaving three weeks after the hurricane. I also conducted follow-up interviews at key points—after Walter and Jennifer returned from subsequent trips, one year after Hurricane Maria, and when they were about to graduate as a cohort of students. This chapter pulls from all of these interviews as well as their frequent social media updates. I followed their work for two years, from late 2017 through till when production concluded.

Hurricanes Maria, Irma, Jose, Dorian, Michael, Sandy, Katrina, and Many More

Epic hurricanes are regularly moving through the Caribbean and the Atlantic coast of the United States, causing billions of dollars in damage and thousands upon thousands of deaths. Scientists are finding clear connections between climate change and these epic weather events. While changing government policies, creating hurricane prediction models, and improving disaster relief programs aim for big picture solutions, are there ways in which a community can adapt quickly?

It is widely accepted and understood that fossil fuel power production is a major cause of climate change. What the scientific community is now seeing is that climate change is causing more extreme weather. Other researchers are quantifying the damage that extreme weather wreaks on our homes, businesses, and communities. For example, a 2020 study published in Scientific Reports found that hurricanes caused 9 out of 10 major power outages in the U.S. and noted that *hurricanes were the leading cause of power outages*, responsible for billions of dollars of damage.[4]

While researchers also are finding better ways to forecast hurricanes, which will allow people more time to prepare and even evacuate, forecasting a hurricane cannot guarantee that there will be a way to protect oneself from it. Nor to preserve critical infrastructure like electrical grids, as that kind of preparation typically must be built into the design of the infrastructure itself.

One way that climate change is making hurricanes worse is by warming the oceans. Hurricanes are a regular part of weather patterns, particularly in the Caribbean. The mega-storms that have moved through the Atlantic in recent years—like Hurricanes Irma, Maria, Sandy, and many others—are far more severe than in the past. Part of their severity is from extreme amounts of rainfall. There is a large body of scientific research collecting on the relationship between climate change and extreme weather.[5]

Severe hurricanes also hit in the Pacific and throughout the world. Because this chapter is based on activities in Puerto Rico, the research and related references are to Atlantic communities, although many of these issues apply to islands and coastal areas in the Pacific.

In a brief summary of this science, hurricanes are caused when warm air currents, loaded with water meet colder air. Air currents travel around the globe at varying layers in the atmosphere, and to some extent, are measurable and consistent, which is what makes it possible to predict hurricanes. But air currents and atmospheric pressure change, and as shown by storms in recent years, hurricanes also change in size, direction, and strength and are not 100 percent predictable. Therefore, evacuating is not always feasible; for example, if you live on a small island, and the hurricane will affect the entire island, evacuation is not a practical option.

The warmer the oceans, the faster the hurricanes become. Warm water allows hurricanes to pick up speed, and these storms create more water vapor in the air. The longer a hurricane travels, the more it picks up speed and ferocity. The more water vapor in the air, the more a hurricane will dump that water when it hits a cooler mass of air, land, or water that stops it. Cold water slows hurricanes significantly. That is why traditionally, hurricanes did not travel as far up the Atlantic coast of the United States, as they ran out of steam when they entered cooler oceans. But with abnormally warm temperatures on land, warming oceans, and air loaded with more water vapor, hurricanes are traveling further, picking up more speed and causing more rain.[6]

When did epic hurricanes begin? And when did the hurricanes become characterized as the problem? In the Caribbean and the Atlantic coast of the United States, hurricanes are as natural a part of weather as blizzards are in northern climates. The problem is not the weather, but our response to and preparation for the weather we know will come. When once-in-a-lifetime hurricanes start becoming annual events, it is only logical to assume that they will continue. Hurricane Katrina, in 2005, was an indication of how deadly a hurricane could be, particularly for low-income communities who might not be able to evacuate or who live in low-lying areas. Every person I spoke with in Puerto Rico, or with Solar Libre, was not thinking about *if* there will be another once-in-a-lifetime event, but about *when*.

Elected leaders from Caribbean nations have testified to legislators about the consistent damage, cost, deaths, and destruction. They have requested resources to build improved infrastructure that will withstand these regular and predictable events. For example, in the first few months after Hurricane Maria, as rebuilding and disaster relief efforts were underway, Governor Mapp of the U.S. Virgin Islands testified to the Senate Committee on Energy and Natural Resources that they have *rebuilt their power grid 5 times* after it was destroyed by hurricanes: "Consider our power distribution network which Irma and Maria destroyed: While we are optimistic that power will be nearly fully restored by Christmas, this will be the fifth time the federal government is paying to rebuild the power distribution system in the U.S. Virgin Islands."[7]

If disaster relief funding pays to rebuild grids over and over again, perhaps that funding could be directed to building a transformed system that was designed based on best practices in withstanding storms, eliminating the need to rebuild continually. Groups like Solar Libre are showing how we can do this.

¡SOLAR LIBRE!—FREEDOM THROUGH SUN

Isabela, a sleepy beach town on Puerto Rico's west side, is home to ¡Solar Libre! A nonprofit begun in the days after Hurricane Maria. Solar Libre started as a humanitarian effort to bring donated solar panels, racks, and brigades of volunteer solar installers to the island. Within a few months, it had started building small, emergency sized solar arrays on the rooftops of nonprofits and community centers.

One year after Hurricane Maria, Solar Libre had built 100 of these small clean energy systems. As they entered their second year, their work attracted a funder who provided enough support to expand and launch an intensive, hands-on solar installer training program. This training program offered industry-standard academic and technical education to island residents, focusing on ensuring the student cohorts were at least 50 percent women. This approach allowed island residents to be part of the growing solar industry on the island and also created a way to expand solar installation training so that there would be more people to conduct this work, maintain systems, repair them when the next hurricane hits. Ideally, these trainees would be part of designing systems optimized for island environments.

In 2019, the Clinton Global Initiative recognized Solar Libre for creating the first solar training program focused on addressing the gender divide in clean energy. Since they began, they have taken the working model that started in the Rockaways neighborhood in New York City and developed it into an organization that is putting down roots. This chapter follows how they got there.

Emergency Power in the Rockaways

Jennifer Bolstad and Walter Meyer are a landscape architect and urban designer team who founded Local Office Landscape and Urban Design, based in Brooklyn, New York. Their work rebuilding with solar after hurricanes began well before Hurricane Maria—they had previously designed and installed emergency solar power systems after 2012's Hurricane Sandy to address the lengthy power outage in New York City's Rockaways, which is located in Brooklyn and Queens on the Atlantic Ocean. The same conditions created that power outage: the grid was blown down throughout an island. Brooklyn and neighboring Queens are located on Long Island, which many forget is still an island, separated from the mainland by bridges and tunnels. Because the Rockaways are also on an island, fuel trucks could not get over bridges or through tunnels after Hurricane Sandy, causing serious fuel shortages. Without fuel, other emergency supplies like generators are useless.

Jennifer Bolstad describes how they began in the Rockaways:

After Sandy hit the Rockaways, we had 12 of our friends camped out at our house in Brooklyn for what was supposed to be a one or two-day hurricane party. Very quickly, it became obvious that they weren't going home anytime soon. We had a lot of smart, capable people with a lot of pent-up energy . . .

that's what we could do. We know how to build small-scale solar generators. No one had power. No one could even use their cell phone and call their families and say, "hey, I'm still alive." Nobody could plug in a power tool and get to work on rebuilding.

The solar power systems dispersed in the Rockaways were tiny, hand-built solar generators about the size of a shopping cart. I asked what the actual distribution was like and whether they worked through an organization or city office. Jennifer shared their grassroots beginning, which focused on meeting immediate needs with immediate action:

We built these really small solar generators together and started dropping them on blocks in the Rockaways where people had kind of de facto started congregating to share resources and knowledge. That scaled up to helping businesses get some solar energy that allowed them to keep their operations and their communications going after dark. [Hurricane] Sandy was late in the fall, and it got dark at about 4:30 in the afternoon. [The solar generators] expanded their capacity . . . giving them enough power to keep their lights on their communications going, and to grow people's capacity to help and be helped.

The impact of these small systems was significant, however, and they provided desperately needed mobile stations for solar electrical generation. However, while this was an immediate start, it clearly could not sustain for very long simply due to limited resources. Jennifer continued, discussing how their initial success attracted funding from New York City:

That effort then scaled up. We got a government grant to install $3 million worth of alternative energy systems to businesses throughout the Rockaways.[8] *That was the progression . . . from this kind of scrappy, you know . . . building panels on our living room floor level . . . to formalizing installations in businesses that are permanent, that can help them to stay in business through the next disturbance, and also help them day to day, to cut down their power bills. Hopefully, they can recoup some of the economic losses of Sandy, which will help create more jobs in the community."*

Jennifer and Walter's vision and leadership in the Rockaways received recognition by President Obama as "Champions of Change."[9] Jennifer shared how they chose the businesses that received the small solar power systems:

Seeing that level of growth with that project, and all of the ripple effects that it had, inspired us to start from that level in Puerto Rico. To go in and say, "what are the businesses that are acting as the critical community hubs? How can we energize them?" So that we can help them help other people, to begin to find their own ability to rebuild.

Applying this successful model of small-scale, easy-to-distribute solar generators after Hurricane Maria shows how quickly equipment can be deployed to meet immediate needs. Unfortunately, while humanitarian efforts after hurricanes and other disasters can be fast and effective, they can also be slow and characterized by delays. In addition, sometimes relief efforts distribute technology that does not help, or use other ineffective approaches. In light of the success of this effort, further research on how disaster relief organizations are using community-based solar generation systems would be valuable to expand efforts like this.

Rooftop by Rooftop, Phase by Phase

When Hurricane Maria destroyed Puerto Rico's power grid, Walter and Jennifer put out a call for action to their community. While they live in Brooklyn, they spend a significant amount of time each year in Puerto Rico, where Walter's father, Thomas, has lived for many years. With an existing model for small solar adapters and a network of people who worked in clean energy, they were positioned to quickly mobilize technology and volunteers to do the same thing in Puerto Rico. Support flowed in from friends, colleagues, state officials, and nonprofits groups in the form of donated solar panels, volunteer installers, and some donated flights and other types of transportation.

A better-designed power system starts with a better-designed plan. Rather than talking about relief, from the first time I spoke with Walter and Jennifer, they sketched out plans that included training island residents to be solar installers and creating infrastructure so isolated rural communities could survive storms without needing relief organizations.

In a November 2017 interview, Walter sketched out the phases of their plan:

> *Phase one of Solar Libre is a humanitarian phase. Its focus is triage, small scale systems in the sub-10 kilowatt range, to energize places without power like community centers or other groups serving some kind of community functions, such as food kitchens. Phase two, the green jobs phase, has already begun. We've had three sites installed as part of phase two, and we are now going to scale it up and target a hundred sites.*

He continued by addressing how simply rebuilding the previous grid system does not make sense, and like Governor Mapp, he sees the logic of building a new and better type of infrastructure:

> When I see the grid being rebuilt, I'm of the mode that this should be a hundred percent solar. And I'm not saying that because I'm inspired by it, or it's visionary. It's actually economically conservative to do that. It's like the most conservative thing you can do is build solar in Puerto Rico. Everything else is a risk. Because you're just going to build the same grid back. And, in 10 or 20 years, another storm will come through and knock it down. They keep saying we have not had a storm like this in 90 years. It does not mean 50, 90 years from now, we will get another one. We know with the statistical effects of climate change that we are getting more intense storms more frequently. The grid creates vulnerability. We shouldn't be building the grid, but we are.

The grid is more than just a piece of technology—it is everyone's connection to their family members in other places. It is the ability to use medical equipment, open for business, go to school and work, and live free of the mental stress of worrying about another loss of power. Walter shared how he saw the human impact:

> For every single line that I see knocked down, every single power pole that's knocked down that I see being built back up, that is enough to take two families, one or two families off the grid, each of those power poles. Each of those things for miles and miles. That's a hundred thousand families who don't have to wait for the utility to build the grid back. That's a symbolic piece that is connected to an economic multiplier. You can't help but think about how each power pole is one or two families, who would not have to pay power bills, and that would free them economically as well.

But without the equipment physically in Puerto Rico, no phases could begin. No design innovation could be built. One of the most significant hurdles in clean energy integration is simply the cost of equipment: not only solar panels but also the racking to mount the solar panels, inverters, wiring, and batteries. The human labor to install all this equipment is another cost.

In the case of an island, shipping is an additional hidden and considerable cost. The donated solar panels were gathering in the Brooklyn parking lot behind Local Office Landscape Architecture. The first day I filmed, volunteers were unloading and stacking donated panels, and several people worked the phones to raise more donated equipment. Walter was on the phone with the office of New York Governor Andrew Cuomo, trying to arrange transportation to the island for literal tons of equipment. This transportation would prove to be one of the most challenging parts of this whole operation.

Getting there is Half the Battle—the Jones Act

Eventually, Solar Libre's network of supporters would grow to include large nonprofits like the Hispanic Federation and the Clinton Global Initiative, along with Governor Cuomo's office and many solar and clean energy companies who donated equipment. But even prominent and connected individuals such as the New York Governor had a hard time securing a flight in the first few months after the hurricane. Transporting thousands of pounds of solar equipment, mostly located in New York state, to an island 1,500 miles away in the middle of the Caribbean would be logistically challenging in any scenario. Walter, Jennifer, and Thomas were navigating this after a major hurricane and during an island-wide blackout. The island's communications systems were nearly completely severed. While you could rent cars and drive, gasoline shortages severely limited the transportation infrastructure, as did damaged destroyed roads and flooding. Government relief groups were calling on any viable trucking to be deployed to send food, water, and emergency supplies.

Some airports on the island were closed entirely, and others did not have power for weeks. The ones that were open had very few flights going in and out. It was unclear when normal airline operations would resume, as even the airports were experiencing power outages.

To get a few tons of solar panels to an island in these conditions could also have been done using ships—except for an archaic law called the Jones Act. This act governs all shipping in the U.S. and became a serious complicating factor. Any discussion of post-hurricane Puerto Rico would be incomplete without a section on the Jones Act and its inhibiting influence, which exacerbated challenges during the relief efforts after Hurricane Maria.

The Jones Act is a 1920 federal law governing shipping in the United States and mandates all goods shipped between U.S. ports must be transported on ships built, owned, and operated by U.S. citizens or permanent residents. As a U.S. territory, the Jones Act governs Puerto Rico. This might not be such a tremendous restriction if the shipbuilding industry in the U.S. was thriving, but it is not. U.S. shipping has waned in light of much cheaper shipping from many other countries. As a result, the number of U.S. ships that meet the criteria of the Jones Act has dwindled. The result is that any industry that needs to ship goods to Puerto Rico, Hawaii, the U.S. Virgin Islands, and any other island governed by U.S. laws is dependent on a minimal number of ships out of the global maritime fleet.

In navigating all of the donated equipment to Puerto Rico, the Jones Act impeded Solar Libre's team many times. First, the solar panels were going to be flown on a plane from New York, then that fell through, and they were trucked to Florida to see if they could ship from there, but due to a very limited number of vessels, it was impossible to find anything. For the first several weeks of their humanitarian effort, Walter and Jennifer were on a daily quest to find a ship or plane that could take their cargo.

Walter describes how this century-old law throttled their immediate efforts:

The Jones Act . . . is good for the mainland to have American ships doing domestic deliveries. But it is terrible for the discontiguous states and territories. . . . If I could have used an international ship, I could have had [solar panels] shipped internationally from China directly [to Puerto Rico] with solar panels at wholesale prices, and cheaper. And have it there quickly. But because there are only a few Jones Act-certified ships, we were limited. They sail twice a week out of Jacksonville, and that is all you get. And they are expensive. Shipping companies don't want to buy new ships because these expensive American ships create overhead for the company. So they maintain older ships that are certified by the Jones Act, and they end up cutting corners to try to compete with other markets. Instead of being a month or two for our gear to get clear from the port, it could have been there in weeks by ship. And, the cost is lower to ship our material on a cargo ship than it is to fly.

The challenges of the Jones Act were not limited to Solar Libre and the development of solar infrastructure. In November 2017—less than two months after Hurricane Maria—then-Governor of Puerto Rico, Ricardo Rosselló, spoke to the Committee on Natural Resources during a Congressional oversight hearing. His topic was the crushing effect that the Jones Act has on Puerto Rico's economy: "Under the Jones Act, Puerto Rico must pay at least double the cost to import goods and supplies from the U.S. mainland compared to neighboring islands, a burden that the island cannot afford."[10] As the hearing proceedings debated the impact the Jones Act has had on the island's economy, Rosselló cited a University of Puerto Rico study that quantified the economic impact of the Jones Act:

A 2010 University of Puerto Rico study determined that the island loses approximately $537 million per year due to the Jones Act. Under the Jones Act, Puerto Rico must pay at least double the cost to import goods and supplies from the U.S. mainland compared with neighboring islands, a burden that the island cannot afford.[11]

After Hurricane Maria, the federal government waived the Jones Act for 10 days, beginning in late September.[12,13] This provided a measure of immediate relief but was not enough. Interestingly, while there had been previous waivers of the Jones Act after hurricanes,[14] the federal government did not do this immediately after Hurricane Maria, and in fact reportedly only took this critical action after Governor Rosselló and many others pressured for it:

> The government issued a short-term Jones Act waiver after Harvey and Irma, but initially hesitated after Maria. It said the earlier waiver was necessary because, without tankers of fuel reaching the Gulf Coast, gas prices would have skyrocketed. . . . But the president seems to have bowed to the political reality that issuing a waiver was politically popular.[15]

While the 10-day waiver provided some additional shipping to allow humanitarian aid supplies like food, water, and gas to reach the island in the first two weeks after the hurricane, the rebuilding of the energy systems would take far, far longer than ten days. As would rebuilding homes, roads, schools, offices, and all the other structures destroyed by the hurricane. The Jones Act waiver would have expired by the time the Solar Libre team navigated their supplies of donated inverters, solar panels, and other equipment from New York, Florida, and other states to a port. They would once again have to use expensive and infrequent Jones Act ships, or even more expensive planes, which were in extremely high demand.

On The Ground Operations

Solar Libre's headquarters were Las Dunas Guest House, a bed and breakfast in Isabela, owned by Thomas Meyer. Tom also served as a liaison in Puerto Rico for all the community groups and managed the highly complex delivery of all the donated equipment. Leading Solar Libre operations meant maneuvering through formidable obstacles like finding space for volunteers for sleeping, drinking water, and food. It also meant managing donated equipment once it arrived on the island.

On November 11, 2017, Solar Libre posted the most beautiful video on their Facebook page: a FedEx plane landing in the airport and crews rolling stacks of solar panels off in plastic-wrapped bundles. With the help of the Hispanic Federation and Ubiquiti Networks, they had finally secured a flight to take the first small shipment of the donated panels to the island.

This was a fantastic development—however, it came with new complications. There were starting to be reports of confiscations of equipment at airports. Just as Walter and Jennifer diligently pursued flights and ships, Thomas had figured out what to do with everything when it arrived. Once on the island, they would have to find a way to transport the gear to Las Dunas. Then, they had to get it to rural communities in the mountains, despite washed out roads, gasoline shortages, not possessing a large truck, and not having a safe place to store their very valuable and sizable pile of equipment.

One morning at an Isabela bakery, Tom ran into another local resident who worked at the airport, who told Tom about the possible confiscation of equipment upon arrival by the U.S. Army Corps of Engineers or FEMA. The press documented these rumors of equipment confiscations at ports. The following January, The Intercept reported that federal agents had to intervene and redistribute much-needed rebuilding supplies that were held in a PREPA-controlled warehouse, which the U.S. Army Corps of Engineers confirmed to Intercept reporters.[16] At this time, in November, it was not certain what would happen. However, with their team of volunteer installers ready to work and after finally securing shipping, the last thing they needed was to have their equipment confiscated.

Thomas shared how a unique set of circumstances helped this situation:

We got 2,000 solar panels that were going to show up in the nearby airport. The only functioning means of communication here is that everybody would go to the local bakery, and they would sit around and talk. And trade information, [like . . .] I'm from this humanitarian group and if you need this, etc. So there was a pooling of resources, and we quite by chance ran into the tarmac director at this bakery. He told us our chance of actually getting all those products—say, maybe a half a million dollars worth of products flying into the tarmac in Aguadilla airport—he said, you've got about a 50 % chance. If the army doesn't redirect your airplane, the national guard will. Or the local Puerto Rican Port Authority. We just fortuitously ran into him, and he said, "you can count on me." He says, "I'll put it in hanger number five, and I'll put it under lock and key." And—a half a year later—the panels are still there. And to this day, we have been taking those panels on brigades up into the mountains. We have probably done a retail of millions of dollars in solar panel installations for underserved communities in a shredded mountain range.

After several months of tackling this non-stop, the panels arrived and were safely stored under lock and key due to Thomas's fortuitous meeting at the bakery. By now, they had also found a truck to take their volunteer crews to the mountains. In addition, the gas stations were beginning to re-open, although roads were notoriously difficult to navigate between floods and destroyed bridges.

On December 9, 2017, they posted another gorgeous video: a drone shot of Walter and a few of the volunteer installers lifting solar panels to the top of an orphanage in Adjuntas, a community located deep within the mountains. They installed other systems at an array of nonprofits and small businesses that had taken on community center-type roles. They delivered a massive mobile solar generator to the hospital in Isabela for use in the emergency room.

In the coming months, Solar Libre would roll out over 100 of these installations by the 1year anniversary of the hurricane. The recipients were organizations serving the disabled, organic farms, a maternal health clinic, small bakeries, and dozens of other rural facilities that were helping the community. Each of these is now ready with a solar-powered, stand-alone energy system, which provides technical relief for the next time a hurricane comes along and mental relief in knowing they will have some type of power. Walter, Jennifer, and Thomas had, by the end of 2017, got through the bottleneck and implemented their Phase One.

TRAINING PROGRAM FOR ISLAND RESIDENTS

The first phase of Solar Libre's operations focused on reaching the small mountain towns—the rural, difficult-to-reach communities located throughout the center of the island. Located along winding roads, these towns were the last to have power restored after Hurricane Maria and the first to lose it in storms. The deep mountain valleys and lush vegetation meant they were likely to lose power again when the inevitable next hurricane came along. These towns are ideal locations for battery-connected solar power systems that can function off the grid.

Intending to do far more than just donate equipment, Walter, Jennifer, and Thomas began the next phase: green jobs. After meeting the immediate need to provide electricity, they looked to address underlying issues and create real sustainability by designing a technical and human system that island residents can maintain.

An underlying and extremely influential issue was that the economy of Puerto Rico had been failing for many years, meaning there were fewer and fewer job opportunities. Before Hurricane Maria, people had been leaving the island to find satisfying work. As communications and transportation were restored after the hurricane, people started leaving the island in even larger numbers, referred to as the "exodus," and heading to Florida, New York, and other parts of the mainland to find jobs and continue their education.

In this environment, Solar Libre started a training program to teach island residents how to become solar installers. Their training program was designed for graduates to pass the North American Board of Certified Energy Practitioners (NABCEP) certification test, the highest level of solar installer certification possible in the U.S. They also focused on recruiting young women as a way to address the gender divide in solar in particular and clean energy in general.

By September 20, 2018, the 1-year anniversary of Hurricane Maria, Solar Libre had installed 100 emergency solar power systems. They also won funding for the training program from the foundation of NFL star Victor Cruz. In addition, they had recruited their first class of apprentices, which was 60 percent women.

The training program prepared the apprentices in a rigorous training program that included a specified number of hours working in supervised installations, as well as being able to conduct unsupervised installations of solar panels, inverters, and batteries. This prepared students to pass the NABCEP certification test and get jobs in the industry. Their trainer, Robert Wylie Hyde, was one of the original volunteers in that Brooklyn parking lot in October 2017; he had moved to the island to become the lead instructor for the training program.

Jennifer describes the impact of these small emergency sized systems in daily life:

"One of the places that we were able to energize was a woman-owned bakery. The owner had started this bakery about a decade ago. She felt like her community was fairly isolated even long before [Hurricane] *Maria . . . that people needed a place they could go every day and get their bread, their coffee, and their basic supplies that was within the community. In that spirit, it had grown to be a fixture of her community, and she was still running it. She had a generator so that she could keep her refrigerator of milk cold and at least bake some bread and serve coffee every day. The skeletal parts of that business were still operating, and it was the only beacon within a really isolated community, located on a lake with really steep cliffs. We saw houses that had slid down the cliff into the lake. It was still very disconnected in terms of the roads being washed out and was very dangerous to get around. Her business was one of the things that lent some hope and sanity and normalcy to people living there. She was in the back, and I overheard her . . . she was praying to be worthy of being the one person that was getting help out of her whole community where everybody had so much need. It was so touching.*

The training program consisted of classroom education from electrical engineering and energy production to construction of a solar power system. It also addressed educating recipients on the best use of their systems, which were adequate to run most, but not all, standard household appliances. For example, fans, clocks, lights, and computers were okay, as were refrigerators, but air conditioners and clothes dryers required too much power to be effective on systems of this size (see "Living with Solar" section below for more detail).

My final shooting day with the Solar Libre folks was in March 2019. The first cohort of student apprentices was doing their first unsupervised solar installation on the roof of an organization called MAVI Arecibo. MAVI provided services to the disabled, and the directors shared how their members fared after Hurricane Maria shut down their operations altogether. A flood had gone through Arecibo, bringing debris and mud down from the mountains, and scattering all of it through the streets. MAVI staff had no way of finding out or contacting people who were in wheelchairs, who had developmental disabilities, or were autistic. These were the people who came regularly to MAVI's office for programs and classes. They described how members who were autistic, for example, would have used landmarks like signs, homes, and plants to independently find their way to MAVI's office. With everything washed away or covered in piles of debris, previously independent people could not navigate this landscape and were confined to their homes. Without any communications, MAVI staff had no way to reach out to their constituency. MAVI's ability to function again after any type of storm or disturbance would support the reintegration of people with disabilities back to independent life.

The Solar Libre crew installed the solar panels at MAVI over the course of two days. The basic steps for the crew were: measuring and attaching the racking, connecting the wiring, attaching the panels, weaving the wiring across the roof and down to the utility room, then attaching it to the battery and other guts of the power system.

MAVI had chosen to have a more extensive solar array installed so they would have the option to operate off-grid on their battery when the power failed again. This installation was done entirely by the Solar Libre training cohort, with the instructor and an independent electrician checking their work. The island's safety regulations require the review by an independent electrician.

After two days in the sun, the training cohort passed. In the next few months, they continued practicing their skills on other installations, developing Solar Libre's reputation as well as their own knowledge and the island's resilience.

Living with Solar

Two of the student installers explained how it was important for recipients to understand the limits of a smaller, battery-connected system. While the solar power systems produce plenty of energy, certain household appliances, such as clothes dryers and air conditioners, use a tremendous amount of that energy and should be avoided or at least minimized. A household can function normally for most other items—lights, refrigerators, computers, television—even on an off-grid battery-operated system. The MAVI system had a battery with a small indicator to show how much power was in it. If it runs low, a household turns off unnecessary appliances. When the sun rises again, the battery will charge.

Systems with batteries, which have been prohibitively expensive but are starting to become affordable, have become very popular in Puerto Rico after Hurricane Maria. Systems with no battery run slightly differently. If the main grid goes down, and the system is wired only to work when the grid runs, then all power goes down. Solar Libre's design allowed the solar infrastructure to be separate from the grid, and run the house during the day, maintaining functionality even if the grid goes down and there is no battery. Because of the small size of the systems, they have to watch the use of electricity and avoid power-hungry appliances like air conditioners. Once the sun sets, however, the system will stop producing electricity. Then, of course, it will start again when the sun rises.

Completely off-grid systems do require some attention to the level of the battery. However, it is possible to have a very functional, typical off-grid household with modern electrical appliances. The installers explained the "do's and don'ts": air conditioners, clothes dryers, heavy draw freezers, electric stoves, electric water heaters, and hair dryers are appliances to avoid because they require too much electricity and drain the system faster. On the other hand, appliances like water pumps, ceiling and floor fans, televisions, washing machines, computers, cell and tablet chargers, oxygen machines, lights—all of these operate very well on a battery system.

The installers share these lists with recipients of the systems and explain the limits of these small systems. When asked how people responded to this relatively slight lifestyle change, they said that the recipients of these systems had lived with no power at all after Hurricane Maria. Seeing what it was like to live entirely without electricity, this firsthand experience living with the barest of necessities made people eager for something that would be reliable and affordable. Understanding how to live within the limits of these systems was far preferable to anything that they had experienced after Hurricane Maria.

Surviving by Design

At the time of writing, Solar Libre was seeking funding to support ongoing training and had to shut down the training program during the coronavirus pandemic. Along the way, they had installed an electric carport with a solar panel and had a small electric truck shipped in from the mainland. They continued a robust program of solar installations throughout the western side of Puerto Rico and were recruiting for their 2021 cohort of trainees, again reaching out to young women.

They also received significant press attention. *Glamour* magazine conducted an interview with Hillary Clinton at a maternal health center called Mujeres Ayudando Madres (Centro MAM, or Women Assisting Mothers), which had a Solar Libre-installed power system. The *Glamour* article discussed the solar power system and Solar Libre's work, and featured Solar Libre field manager, Paola Pagán Berrios.

CONCLUSION

Solar Libre's work is tangible, hard-wired proof that we can redesign our grid and homes and create community resilience to survive extreme weather. The large-scale, utility-led initiatives described in other chapters of this book create much larger solar and wind farms, but they are still working on grid resiliency. Many working in energy fields are calling for this kind of transformation, allowing single-family homes to island off the grid and protecting critical sites like hospitals with large-capacity solar generators.

Most importantly, to transform power systems, not only is the large and systemic approach right but also the small and modular approach for they are *both* the right ways.

As Walter put it: "We're *working from the bottom up, and they're coming from the top down, but we'll meet in the middle and hopefully get all of Puerto Rico to be fully renewable*" (Emphasis mine).

As clean energy systems become more and more ubiquitous, it is becoming far more apparent how to design homes and businesses to function better during and after storms, and how to ensure whole communities will have safe, reliable power. Batteries are a key aspect of these systems. With prices steadily dropping and availability steadily increasing, we will see a lot more of these small-sized solar-and-batter power systems, particularly in hurricane-prone areas.

Seeing these small systems as functional, accessible, everyday parts of a home's infrastructure goes a long way toward undoing outdated narratives like "solar doesn't work." These small systems not only transform the technology of our electrical system, but they also transform our attitudes in ways that will help us deal with the massive challenges of climate change. Jennifer summed up how this attitude shift is equally essential along with the technological transformation:

> *I want to make a pitch for maintaining optimism in the face of global climate change. The only way that I can continue to work in the space of climate change is to foolishly or otherwise maintain that optimism. Something as simple as lifting three solar panels onto the roof of a bakery, then wiring it down into some batteries, plugging it into someone's panel, and then saying, now I can run X, Y, and Z that I couldn't do before. There's a lot of very tangible optimism in that. I think holding onto that optimism is a necessary objective for anyone who's*

working in the space of climate change. Because once it gets to be pure gloom and doom, then we all throw up our hands and say, "too bad there's nothing I can do." And that stops our progress.

No matter when the next hurricane hits Puerto Rico and the rest of the Caribbean islands, Solar Libre made it a safer place. Starting with handmade solar kits in the Rockaways, they created a working model of better design, solar for emergency use, and solar as a means to foster local economic development.

NOTES

1. Englund, Will. "The Grid's Big Looming Problem: Getting Power to Where It's Needed." *Washington Post*, June 29, 2021. https://www.washingtonpost.com/business/2021/06/29/power-grid-problems/.

2. Clark, Jeremy. "Emerging Best Practices for Utility Grid Hardening." *Utility Dive*. November 5, 2018.

3. Eyocko, Stephanie. "Utilities Need to Harden the Grid as They Green It. Consumers Aren't Ready for the Cost." *Utility Dive*. February 26, 2021.

4. Alemazkoor, N., B., Rachunok, D.R. Chavas, et al. "Hurricane-induced Power Outage Risk Under Climate Change Is Primarily Driven by the Uncertainty in Projections of Future Hurricane Frequency." *Science Report*s 10, 15270 (2020). doi: 10.1038/s41598-020-72207-z

5. Francis, Jennifer. "Yes, Climate Change Is Making Severe Weather Worse." *Scientific American* June 1, 2019 https://www.scientificamerican.com/article/yes-climate-change-is-making-severe-weather-worse/

6. Marsooli, Reza, and Ning Lin. "Impacts of Climate Change on Hurricane Flood Hazards in Jamaica Bay, New York." *Climatic Change,* November 26, 2020. doi: 10.1007/s10584-020-02932-x

7. Mapp, Kenneth. "Written Testimony of Governor Kenneth E. Mapp of the United States Virgin Islands Before the Senate Committee on Energy and Natural Resources." *U.S. House of Representatives, 105th Congress, 1st Session, Committee On Natural Resources*, Tuesday, November 14, 2017. Full Hearing Text: https://www.govinfo.gov/content/pkg/CHRG-115hhrg27587/html/CHRG-115hhrg27587.htm Gov. Mapp's Written Statement: https://www.energy.senate.gov/services/files/A2538A49-2953-4BA1-8C94-0807E62050A5

8. Eyoko, Stephani. "Utilities Need to Harden the Grid as They Green It. Consumers Aren't Ready for the Cost." *Utility Dive*. February 26, 2021.

9. Obama, Barack. "A Shining Light - Summary: Walter Meyer Is Being Honored as a Champion of Change for the Leadership He Demonstrated in His Involvement in Response and Recovery Efforts Following Hurricane Sandy." *The White House blog*, June 19, 2013. https://obamawhitehouse.archives.gov/blog/2013/06/19/shining-light

10. Rosselló, Ricardo. "The Need for Transparent Financial Accountability in Territories' Disaster Recovery Efforts." Congressional Testimony, Committee on Natural Resources, November 14, 2017. https://www.govinfo.gov/content/pkg/CHRG-115hhrg27587/html/CHRG-115hhrg27587.htm

11. Rosselló, Ricardo. same.

12. Rosa-Aquino, Paola. "What Will a Jones Act Waiver Mean for Puerto Rico's 100% Renewable Energy Goal?" *Grist*, April 26, 2019. https://grist.org/article/what-will-a-jones-act-waiver-mean-for-puerto-ricos-100-renewable-energy-goal/

13. Homeland Security Department. "Waiver of Compliance with Navigation Laws; Hurricane Maria." *Federal Register*, October 4, 2017. https://www.federalregister.gov/documents/2017/10/04/2017-21283/waiver-of-compliance-with-navigation-laws-hurricane-maria

14. Homeland Security Department. "Waiver of Compliance with Navigation Laws; Hurricanes Harvey and Irma." *Federal Register*, September 14, 2017. https://www.federalregister.gov/documents/2017/09/14/2017-19523/waiver-of-compliance-with-navigation-laws-hurricanes-harvey-and-irma

15. Graham, David. "Is the Jones Act Waiver All Politics?" *The Atlantic*. 2017. https://www.theatlantic.com/politics/archive/2017/09/jones-act-waiver-puerto-rico-trump/541398/.

16. Aronoff, Kate. "Armed Federal Agents Enter Warehouse in Puerto Rico to Seize Hoarded Electric Equipment." *The Intercept*. 2018. https://theintercept.com/2018/01/10/puerto-rico-electricity-prepa-hurricane-maria/.

Chapter 8

La Riviera, Puerto Rico
Resilient Infrastructure

INTRODUCTION

La Riviera, Corozal is located high in the mountains south of San Juan. During the Hurricane Maria grid failure, not only did they lose their electricity, but they also lost the use of their water pumps. Drilled hundreds of feet into the rugged mountains and powered by the grid, the pumps stopped working. Not having electricity is dangerous and difficult. Not having water is deadly. The community passed a hat to purchase a generator and then kept passing it to cover the fuel costs. They made regular fuel runs, navigating mountain roads that had been damaged or washed away by the hurricane, sometimes having to drive miles in a different direction because no one had yet repaired a bridge that had been destroyed. For months, they purchased enough gas for the generator to fill a large cistern perched higher up a hill. Conserving and rationing, the entire community used the water in that cistern until they could get more gas and pump again. With gas shortages everywhere on the island, there was a perpetual looming fear that this fragile system would fail as well.

La Riviera residents had waited for the Puerto Rican utility to bring power lines back to their community and restore their water. At the time I was in La Riviera, the next hurricane season was just 3 months away. The potential to lose this limited access to roads and their inadequate fuel supply was on everyone's mind. Anxiety had overshadowed every day as the potential for more hurricanes grew closer. Finally, they connected with the nonprofit Water Mission, an international organization that builds safe water and sanitation systems in developing countries and disaster areas. Many communities in Puerto Rico were in the same situation as La Riviera residents—living without functional water systems for months. By the time we were filming,

Water Mission had built several battery-connected, solar-powered water pumps throughout the rural areas of Puerto Rico.

Water Mission builds solar-powered water pumps for communities in need, typically in countries far more distressed and impoverished than Puerto Rico. This was an extraordinary situation, and they had partnered with Blue Planet Energy, a Honolulu-based company that engineers energy storage systems designed for off-grid systems, particularly in island environments. Blue Planet Energy is a private business and also a foundation, both with the mission of eliminating carbon-based fuels and powering the world with clean, affordable energy.

In the first year after Hurricane Maria, Water Mission installed dozens of donated battery-and-solar systems in mountain towns throughout Puerto Rico. All were in areas where people had been waiting for months for water service to be restored, and they completed all of them well before PREPA reconnected the power grid.

CHAPTER SUMMARY

This chapter looks at the connection between water and power and the need for consistent, reliable power for water pumps to run. The role between energy and critical infrastructure is well known in the industry, although it is perhaps lesser-known in public spheres. Seeing this role for clean energy, particularly nimble, modular solar panels, which can be installed in places where other types of energy production cannot fit, expands public concepts of how towns and cities can use it.

This chapter also looks at off-grid battery-attached infrastructure in rural areas. The majority of chapters in this book look at systems that are attached to the grid. In this chapter, the water pump system was designed to go off the grid and function in any kind of weather. This case study of how Water Mission and Blue Planet Energy modeled a resilient solution for La Riviera offers insights into how solar can support cleanly powered, reliable infrastructure everywhere.

METHODOLOGY

I asked Walter Meyer, the co-founder of Solar Libre (see chapter 7), if he knew anyone bringing batteries to Puerto Rico in Hurricane Maria's aftermath. He introduced me to the Blue Planet Energy executives Gregg Murphy and Kyle Bolger. Gregg and Kyle, in turn, introduced me to Mark Baker, Water Mission's Director of Disaster Response.

At their invitation, I went to La Riviera in March 2018 with Water Mission and Blue Planet Energy and interviewed La Riviera community leader Juan Santana. A few weeks later, a Puerto Rico-based producer and cinematographer returned to film the unveiling of the completed system in April 2018; he conducted a follow-up interview with Juan. Additional information in this chapter comes from press releases about the partnership between Water Mission and Blue Planet Energy and industry press on the use of solar and energy storage for disaster response.

OFF-GRID WATER PUMPS

Resiliency

On September 20th, 2017, when the news media worldwide reported on Hurricane Maria hitting Puerto Rico, the massive loss of electricity throughout the island shocked and horrified people everywhere. Social media shares of a photo of the island taken from space and compared to a previous image provided a dramatic visual representation of the entire island in the dark, but especially the mountainous area in the middle of the island. The tiny number of lights that remained were clustered by the coasts, most near the capital city of San Juan.

Across the island, the hurricane washed out roads, caused landslides that tumbled houses, and ripped roofs off homes. It seemed that every semblance of modern life was gone. Among all this destruction, one of the most dangerous things was also the least visible: the loss of clean water. The blackout shut down many water pumps everywhere, particularly in rural areas, and contaminated water supplies in urban ones.

Access to clean water and sanitary conditions is a marker of a civilized country. So much so that one of the United Nations' sustainable development goals is ensuring that all people have access to water and sanitation.[1] Around the world, researchers explore how to improve access to water and modernize agriculture and rural communities, including bringing electricity, without the cost, complications, or unsustainable future of fossil fuel electrification. For example, a 2018 study in India looked at negative aspects of using fossil fuel water pumps for irrigation, including quantifying their contribution to climate change, finding that diesel-burning water pumps added 45 million tons of carbon dioxide to the atmosphere. It then also explored the opportunities and advantages of installing solar photovoltaic water pumps instead.[2] Other studies have also researched the use of solar for water pumps, with studies looking at various topics such as: whether decentralized off-grid solar irrigation systems are effective in poverty reduction, particularly for female heads of household;[3] opportunities for using solar in farming in developing countries with inadequate irrigation systems;[4] and detailing the configurations of off-grid systems to discern the most economical and feasible designs.[5]

This kind of research helps illustrate the close relationship between electricity and water and how solar can have uses that might go beyond the typical rooftop photovoltaic system for powering an individual home. As damage from extreme weather becomes worse, research in sustainable energy needs to prioritize looking at technical and design solutions capable of protecting critical infrastructure like water pumps from the worst impacts of extreme weather. Every year, hurricanes batter communities, knock out power grids, and destroy infrastructure. Combined, blackouts and power outages cost thousands of human lives and billions of dollars in property damage. Inconsistent and unreliable power, and damaged power systems from storms, should not be a given. With energy storage systems becoming more affordable, we now have the capacity to protect communities from losing critical infrastructure like water pumps.

La Riviera, Corozal

Juan Santana is the elected leader of the middle-class La Riviera community, located high in the mountains south of San Juan. After Hurricane Maria, when La Riviera's water pumps stopped working, Juan and his neighbors were left without a means to access drinkable water. Their water pump went hundreds of feet down into the mountain, and while its mechanics were still intact, the pump could not function without electricity.

Neighbors in La Riviera passed a hat and bought a generator. For seven months, they kept contributing money for the community to purchase gas for the generator so that they could pump drinking water. Once a week, someone would navigate the treacherous roads, find an open gas station, and typically wait in a long line for scarce gas. Finally, they would pump just enough water to fill a large hilltop cistern, which would be the entire community's potable water for the week. For seven months, they lived with the deep anxieties—about this system breaking down, about maneuvering the treacherous drives down roads destroyed in the flooding, and about the frequent gas shortages.

Walking us through his neighborhood, houses decorated with Puerto Rican flags and friendly dogs following alongside us, Juan pointed out the other communities throughout the region. The view from La Riviera looked over a massive ravine to the next mountain. There was another mountain after that one, and then another mountain—each peppered with small communities like La Riviera. In the months immediately after Hurricane Maria, Juan told me, PREPA representatives from the Puerto Rican utility (PREPA) visited a few times, always promising that power would be restored soon. Sometimes,

Figure 8.1 **Rooftop Solar Water Pump in La Riviera.** *Source*: Fabrizi, Leandro, 2018.

promises were made that power would come by the next week. Though, it never materialized into anything.

Facing this inadequate government response, the unsustainability of their makeshift water system, and just a few months from the start of hurricane season in June, La Riviera residents undertook reconstruction themselves. Through Christian networks, they connected with Water Mission, a South Carolina-based Christian nonprofit. Water Mission had not previously worked in Puerto Rico, but Maria changed that. Within a few months, the organization opened a field office in Puerto Rico and had created a supply of donated gear and utility-scale battery systems. Soon they formed a partnership with Blue Planet Energy, which specializes in batteries for rugged island environments. Blue Planet contributed many large-capacity batteries to the relief effort.

Mark Baker, who manages Water Missions disaster response programs, described their goals: *"Much of Puerto Rico still lacks access to clean drinking water and electricity. Our mission on the ground in Puerto Rico is to coordinate with the EPA and FEMA to install safe drinking water stations and solar-powered pumping systems to service those that need it most"* (Emphasis mine).[6]

Water Mission worked with FEMA and the EPA, who helped choose the sites with the most need, with a focus on the mountain towns with inaccessible roads and rough terrain. From conversations with Juan and others, it was evident that the mental health toll of the apprehension and dread people were experiencing was another result of the power outage. This anxiety was exacerbated by the expectation that if anything did get repaired by PREPA, it would probably be done inadequately and would likely break down very soon. The solar-powered water pump offered more than just water—it provided peace of mind and a glimmer of security in all of the turmoil and chaos.

People loved living in La Riviera's close community. With a smile, Juan told me he was born in La Riviera, lived there his whole life, and expected to die there. It was clear how dedicated he was to his community. But the fear and uncertainty were, understandably, driving people to leave. The area's population, only 120 families before the hurricane, had fallen to 60 because of the lack of drinking water.[7]

The technical solution: Blue Planet Energy and Water Mission added a 16kWh battery storage system and solar 7kW solar array. On the March 2018 trip, when we were filming, the battery and connections were attached to the water pump. The final piece—the solar panels would be installed as soon as possible. In the months after the hurricane, professional solar installers were booked solid, and Water Mission and La Riviera were just waiting to confirm a date. Finally, one month later, the installation was complete. I arranged with the cinematographer to return and film the solar panels and celebration when they unveiled the system.

Installation time was minimal; it took less than 2 days total. Water availability was immediate. As soon as the system charged up, it went from only a few hours a week, using the generator, to 12 hours per day using the solar power system. The installation was done by a Puerto Rican solar company, New Energy Consultants, providing a local resource to contact if the equipment ever needed maintenance and contributing to the island's solar economy development.

The solar and battery systems eliminated the need for expensive, tenuous, and unnecessary dependence on fossil fuel-powered generators for La Riviera residents and other towns which received these systems. The intangible benefits, like a sense of relief and security, were incalculable.

Many companies came to Puerto Rico after Hurricane Maria to bring clean energy technology to the island. For some, it was about much more than marketing their brand. According to Kyle Bolger, an engineer and trainer with Blue Planet Energy, they want to do more than just make a donation:

> *Training and education are part of Blue Planet Energy's larger mission of powering the world with clean energy combined with long-lasting, reliable, and grid-independent battery storage. Being on the ground in Puerto Rico and speaking with people from communities impacted by Hurricane Maria, we've*

seen firsthand the risk that centralized power systems pose and the hardship they can leave in the wake of a devastating weather event. The Blue Planet Energy team is thrilled to pass on the knowledge and tools for reliable, well-designed off-grid power so that Puerto Ricans can rebuild their communities.

At a celebration for the newly built water-pumping system in April 2018, Juan Santana spoke about how the solar-powered pump not only met the community's day-to-day needs, and also positioned residents to weather the next storm, no matter its intensity. With the solar system activated and connected to the battery and ample time to correct any technical issues that might arise, they will have a durable water system when the next hurricane hits La Riviera.

Larger Scale Impact

By December 2017, Water Mission had invested over $1,000,000 in labor and equipment in Puerto Rico—all donated. Water Mission restored water pumps in over 50 communities and had 30 more planned.[8] These solar-powered water pumps solve the crucial and immediate problem of providing water after a hurricane. On a larger scale, they address a significant barrier to accessing clean energy technologies. While they are highly effective, they are also expensive. Many communities cannot pass the hat enough times to ever afford technology like a large solar panel array with a resilient battery. Nor can they readily navigate challenges like shipping, finding qualified installers, managing the maintenance, and figuring out all the other practical hurdles. Creating these alternative systems in Puerto Rico through donated equipment and skills as part of hurricane relief efforts is an unparalleled way to level the energy playing field. It also opens the door to future innovation, as these front-line communities now have access to cutting-edge technology. Young people growing up in these towns might never learn the misinformation surrounding solar panels; they might develop new ways to adapt this technology and further meet the needs of their unique communities.

The La Riviera site, and others like it, are also microgrids for water. After installation, the system ran entirely on solar power for months. When grid power was finally restored in La Riviera, over a year after Hurricane Maria, the water system was reattached to the grid, with the battery system remaining, and continued to use the solar panels to operate. This type of a system speaks to another design innovation: it is capable of running attached to the grid and off the grid, increasing the flexibility of the system and creating distributed generation throughout the island. When grid power is available, the water pump can run off-grid power. The solar energy it produces goes back into the utility grid. This type of system lowers the community's operating cost, saving La Riviera 75 percent of the previous costs of running the water pumps.⁹ This example shows how solar can be part of resilient infrastructure, potentially save lives, provide security, while also improving overall community economics.

Energy Storage Systems throughout The Island

This battery-and-solar power system was designed specifically for a water pump in a rural area. Many Puerto Ricans turned to backup batteries after the hurricane, purchasing them from Blue Planet Energy and other companies. Residential batteries for energy storage were just starting to be available for consumers, and while their cost had come down, they were considered an expensive item. Many people who installed solar before Hurricane Maria could not also afford to purchase a battery. Other people who bought solar simply did not understand that their solar panels cannot power their house when the grid fails.

One residential site where Blue Planet installed a battery system was Las Dunas guest house in Isabela. Discussed in chapter 7, Las Dunas is also the headquarters for Solar Libre. With volunteers coming in the months after the hurricane, Las Dunas had been operating without electricity, despite trying to launch a humanitarian effort to install solar power on community centers throughout the island. They needed electricity to manage a relief effort and house over a dozen volunteers.

Las Dunas and La Riviera were just two of the many places buying off-grid batteries. By February 2018, "energy storage" was a household term in Puerto Rico, even though solar panels combined with energy storage systems had barely hit mainland markets. This transformation was a fast "citizen science" type learning curve, with communities helping each other understand where to get a battery and how to create a microgrid. The technology is quite simple to understand, and the need was evident. People who previously resisted because of the cost saw why being an early adopter of energy storage was a worthwhile investment. Public understanding advanced to the point that grid defection—leaving the grid entirely, in some cases paying a fine to do so—and interest in completely off-grid systems also rose.¹⁰

To longtime solar advocates, this was a welcome technological advance. Far from La Riviera and in the mountains to the west, Casa Pueblo, in Adjuntas, has been a leader in solar technologies in Puerto Rico for years. Among other developments, they built a solar-powered cinema, a solar-powered radio station, and a solar-powered barbershop, as well as a research and education program. In a 2018 editorial, Casa Pueblo Executive Director Arturo Massol-Deyá stated: "Hurricane María brought an opportunity to move away from a fossil fueldominant system and establish instead a decentralized system that generates energy with clean and renewable sources. This is the path that will bring resilience to Puerto Rico. Citing the many off-grid localized renewable energy production sites now installed on the island, he noted how these systems "are changing the energy landscape of the municipality. But the majority of rural communities are still in need of sustained help."[11]

CONCLUSION

In an anthology about many aspects of life in Puerto Rico after the disaster, Massol-Deyá notes: "In Adjuntas, we showed that alongside the despair, there exists another Puerto Rico, one where diverse, organized communities responded to the emergency with self-management and a collective sense of hope." As in the other chapters of this book, Water Mission, Blue Planet Energy, and La Riviera found that collective sense of hope and self-management. Rising together out of utter disaster, they leapfrogged chasms of technology to install renewable energy systems that immediately impacted daily life. Operating out of a humanitarian effort, they showed that, while despair and dread can plague people's everyday existence, a deeper sense of hope and collective action is more powerful. Their work also speaks to how embracing the transformation of the island's energy system and that collective sense of hope can transform the future of rural neighborhoods.

NOTES

1. United Nations. "Sustainable Development Goals, Goal 6: Ensure Access to Water and Sanitation for All." *UN.org*, accessed December 29, 2020. https://www.un.org/sustainabledevelopment/water-and-sanitation/

2. Pushpendra Kumar Singh Rathore, Shyam Sunder Das, and Durg Singh Chauhan. "Perspectives of Solar Photovoltaic Water Pumping for Irrigation in India." *Energy Strategy Reviews* 22 (2018), 385–395. doi: 10.1016/j.esr.2018.10.00

3. Wong, S. "Decentralised, Off-Grid Solar Pump Irrigation Systems in Developing Countries—Are They Pro-poor, Pro-environment and Pro-women?" In Castro, P., Azul, A., Leal Filho, W., and Azeiteiro, U. (eds.), *Climate Change-Resilient Agriculture and Agroforestry. Climate Change Management.* Springer, Cham.

4. Roblin, Stéphanie. "Solar-powered Irrigation: A Solution to Water Management in Agriculture?" *Renewable Energy Focus*, 17, no. 5 (2016), 205–206, doi: 10.1016/j.ref.2016.08.013.

5. Sawle, Yashwant, S.C. Gupta, and Aashish Kumar Bohre. "Review of Hybrid Renewable Energy Systems with Comparative Analysis of Off-grid Hybrid System." *Renewable and Sustainable Energy Reviews*, 81, no. 2 (2018), 2217–2235. doi: 10.1016/j.rser.2017.06.033.

6. Blue Planet Energy. "Press Release: Blue Planet Energy Joins Puerto Rico Electrification Efforts with Battery Deployments And Trainings." *Pv-magazine-usa.com*, March 2018. https://pv-magazine-usa.com/press-releases/blue-planet-energy-joins-puerto-rico-electrification-efforts-with-battery-deployments-and-trainings/

7. Blue Planet Energy. "Case Study: Water Mission Delivers Clean Water to over 1,200 People with Blue Ion in Puerto Rico." *BluePlanetEnergy.com*, October 30, 2018. https://app.box.com/s/6tq1hwughe90cnif4ijo3s39klax3suc

8. Water Mission. *Watermission.org*, accessed December 29, 2020.

9. Burger, Andrew. "Microgrids Restore Water, Wastewater Services for Isolated Puerto Rico Communities." *MicrogridKnowledge.com*, April 2, 2019. https://microgridknowledge.com/microgrids-water-puerto-rico/

10. Foehringer, Emma. "Grid Defection Is on the Rise in Puerto Rico." *GreenTech Media*, February 16, 2018. https://www.greentechmedia.com/articles/read/grid-defection-on-the-rise-in-puerto-rico

11. Massol-Deyá, Arturo, Jennie C. Stephens, and Jorge L. Colón. "Renewable Energy for Puerto Rico." *Science,* 362, no. 6410 (2018), 7. doi: 10.1126/science.aav5576

Chapter 9

Highland Park, MI
Solar-Powered Streetlights and Community Activism

INTRODUCTION

Highland Park, Michigan is most well-known for two things: 1, at the beginning of the 20th century, as the home of the first Ford Model-T factory; and 2, as the auto industry declined and factories were closed, Highland Park became one of Michigan's poorest zip codes. Now, this tiny city is starting to influence energy legislation that will impact millions of people in southeast Michigan.

With a current population of about 10,000, this small city, located inside the Detroit city limits, is surrounded by Detroit's 665,000 residents, and also the Detroit metropolitan area, which encompasses six countries and has over 5 million residents. Highland Park is home to some of the most creative and effective uses of solar power in the U.S., which are designed to also address issues of equity and economics. Organized by an independently founded non-profit, and engaged in many statewide collaborations, Highland Park is leading the installation of solar-powered streetlights, energy education programs, and is part of larger efforts to create community solar in Michigan.

This chapter explores community-based solutions to systemic environmental justice issues. The non-profit, called Soulardarity, provides education about energy, lighting, and ownership, campaigns to engage their members to participate in statewide energy decisions, and community solar programs including solar streetlights. In a process that began with a handful of solar-powered streetlights, Highland Parkers ultimately forged an entirely new dynamic with one of the largest utilities in the nation, when that utility repossessed their streetlights.

DTE Energy (DTE) serves over 2 million people in Southeast Michigan, is one of the largest utilities in the nation, owned by investors, and traded on the New York Stock Exchange as one of the nation's largest 500 companies. DTE has over 10,000 employees and has reported annual revenue exceeding $12 billion dollars every year for the past 5 years. Soulardarity is a tiny non-profit that began with door-knocking and volunteers meeting in a local cafe and bookstore overflowing with art, clothing, and publications on wide-ranging topics. Over the years, Soulardarity grew to a staff of five full- and part-time workers, and in 2019 they published an annual report thanking about a dozen foundations and donors and detailing their budget of just under $300,000. Yet, despite these extremely limited resources, Soulardarity succeeded in lighting the streets of Highland Park with solar and in engaging residents in learning about energy systems in general, as well as how to integrate solar in whatever way they could.

Taking on a titan-sized utility is not something every small non-profit aspires to do, yet with a fraction of the massive resources that DTE has at its disposal, Soulardarity, along with environmental groups from across Michigan, the state Attorney General, thousands of state residents, and hundreds of state employees, influenced the energy policies that DTE submitted to the state for review and approval at the end of 2019—policies that would govern DTE's actions and rates for the 15 years. Utility policies must be approved by a state energy commission, and are subject to a period of public commentary. This statewide coalition organized people in Highland Park and across the state to submit comments and testimony, and a judge sent DTE's proposed plan back to them, telling them to make more specific plans for solar and wind, along with other suggestions.

This was an enormous triumph that took an entire statewide network to win. For Soulardarity, there is a special impact: after about a decade of organizing, they had introduced an entirely new way to think about energy systems to a neighborhood that was about as alienated from state policy as possible. This new way of thinking was turned into action during this public comment period, as well as through other activities, which produced tangible, and far-reaching results on a policy level. As cities throughout the U.S. plan to integrate new renewable energy sources and expand existing ones, this chapter looks at how Highland Parkers have shown the entire nation an entirely new meaning for the term "energy independence."

CHAPTER SUMMARY

This chapter looks at a community-based initiative to build solar-powered street lights in Highland Park, Michigan after the utility repossessed well over 1,000 streetlights. This initiative started very small—a few people in a cafe—and ultimately ended up impacting statewide renewable energy policy. This chapter explores how a community used clean energy as part of a larger solution to longstanding problems, including looking at how these longstanding problems are rooted in neglect, the loss of jobs, and environmental justice. It dives into how Highland Parkers found their own solutions, from technical fixes to policy/advocacy campaigns, educational programs, and investing resources in their own community.

Another unique theme of this chapter is to introduce the concept of energy democracy as related to the event in Highland Park, and explore how a community can participate in energy systems from production through distribution. To do so, this chapter follows the path taken by Soulardarity, a local non-profit with a mission "to educate, organize, and build people-powered clean energy." Soulardarity centers its practice in the idea of energy democracy, which is being put into practice by many organizations around the world and is central to understanding the rest of this chapter. Energy democracy is becoming the subject of academic research and represents a significant transformation of our thinking about our energy systems. For many clean energy movements, it is becoming a central organizing concept. Soulardarity's definition of energy democracy is: "the idea that the people most impacted by energy decisions should have the greatest say in shaping them."[1] A more detailed discussion of this definition is below.

METHODOLOGY

In looking for unique stories about solar development around the U.S., I came across a list of cities that had stated their commitment to supporting the Paris climate accords. On this list was Detroit. As someone who grew up in Michigan, I knew Detroit as The Motor City, and this intrigued me. How will a city whose claim to fame is the Automotive Capital of the World set out programs to support historic climate accords which limit fossil fuel use and mandate renewable energy transition? I started researching solar, wind, and geothermal projects in the Detroit metropolitan area, and found several initiatives, including a "Detroit 2030 Project" to advance energy efficiency and cleanly-powered buildings amidst the renovations happening throughout Detroit. I also found many media stories about a very unique use of solar: the solar-powered streetlights.

I contacted the non-profit, Soulardarity, whose members and leaders were featured in many of the media stories. I also contacted a Highland Park resident, a historian and journalist by profession, who had recorded video of the streetlights being repossessed in front of his home, posted them on YouTube, and shared them with the media as documentation of what had happened. I shared what my goals were in making this film with everyone and asked if they were interested in being a part of this film. Everyone said yes, and we started planning the interview time.

Soulardarity works with many others to create the streetlight program as well as their other education programs. Their staff—Jackson Koeppel, the co-founder and Executive Director through mid-2021, and Shimekia Nichols, the incoming Executive Director—introduced me to Juan Shannon, who was born and raised in Highland Park, and an entrepreneur leading a new development called Parker Village. Shannon had purchased an abandoned school and is renovating it into an eco-village which would include solar-powered streetlights, a large community garden, other sustainable design capacities like aquaculture and an eco-café, and, eventually, STEM education programs. The Soulardarity staff also introduced me to Ali Dirul and Karanja Famadou, the principals of Ryter Cooperative Industries (RCI), a widely respected local company creating practical yet visionary clean energy projects for homeowners and businesses. RCI created and installed a variety of solar solutions around Metro Detroit, including affordable streetlights in Highland Park, and has been widely recognized for creating practical yet visionary renewable energy solutions for people who typically do not have access to technology like solar panels. I also spoke with the owner of Nandi's Knowledge Café, who was one of Soulardarity's first board members.

Everyone in Highland Park was responsive and generous with their time and knowledge in researching this story. During interviews, each person was asked to share their personal connection to this larger story about a community embracing a very unique way to use solar. This narrative approach allows for an exploration of the many interconnected disciplines that are involved in the transformation of energy systems: engineer, entrepreneur, advocate, elected official, business owner, and consumer. It also provides insight into how a community can transform the ways we *think* about energy as well as how we *use* energy.

Additionally, my research included speaking with several advocates for solar energy and green buildings in surrounding Detroit, including those leading the 2030 District. At a time when Detroit was experiencing significant rebuilding, renovation, and seeing new developments, particularly in Midtown and Downtown, these initiatives introduced concepts of renewable energy and high-efficiency buildings at a key time, helping guide the city towards meeting those Paris climate accord goals. To the degree possible, I have included additional references to other renewable energy developments in surrounding metropolitan Detroit.

TERMINOLOGY

Two Terms Requiring Definitions

Solar-powered Streetlight

If the term "solar-powered streetlight" does not ring a bell, you are not alone. Readers of this book might not have any familiarity with them, have never seen one, and might be wondering about what, exactly, solar-powered street lighting is, as well as the need for it. The basics; a solar-powered streetlight is exactly what the name says: a regular streetlight on a pole, strong enough to light a full parking lot, yard, or decent amount of the street, and powered by solar panels. 2 models were installed in Highland Park and were shown in the film and in figure 9.1.

One, a very modern style installed at Parker Village (an eco-village discussed below) was a high-tech looking design, with solar panels running along the pole from the top to bottom. The other positioned the solar panels at the top in a wing-shaped pattern on either side of the light. The lights have

Figure 9.1 Solar-Powered Streetlight at Parker Village. *Source*: La Rosa, Melanie, 2018.

LED bulbs and batteries, which charge during the day from the solar panels. Because LED bulbs are extremely energy-efficient, the battery provides sufficient power for them to run all night. Some of the streetlights in Highland Park are also "smart lights," and feature Wifi routers and motion detectors, which enable the lights to run at a dim setting initially, and brighten when a car or person walking is detected, which conserves the battery power. These lights are also completely detached from the grid, meaning they will function in blackouts. Once installed, they run for practically no cost, with the primary maintenance being an occasional battery change.

Energy Democracy

"Energy democracy" has several working definitions of the concept and a range of ways to put it into practice. However, all of them have a common theme: transformation of the ownership and economic benefits of energy production, as well as the technology of making it. In practical terms, this means the "decentralized production of electricity"—a somewhat technological way of saying that electricity can—and should—be produced by many smaller production sites. Currently, our energy system creates electricity at large, centrally-located plants like nuclear plants and the natural gas, oil, coal-burning plants most people think of when they hear the term "power plant." Decentralized production refers to this production moving to solar rooftops, solar farms, wind farms, and other types of energy production spread throughout a city or town. It can also mean producing power closer to the point of the energy being used; for example, many factories and companies with large warehouses are installing massive solar rooftops that allow them to cut their energy bills. This is one type of decentralized production.

There is a growing body of academic research on energy democracy. A well-known scholar on energy democracy defines it as: "taking ownership of energy into our own hands, sketching out a world where renewable energy makes it possible to have energy production in our own backyards, at locally owned solar and wind farms, and other smaller sites."[2] This definition, echoed by other organizations and individuals promoting energy democracy, suggests turning to other models of energy production, which are now possible and coming into use in many places, like rooftop solar on homes and solar or wind farms owned by a community entity, typically a cooperative or municipality. Called "community wind" or "community solar," these include careful local siting, allowing one site with the right sized building or available land to produce power for many buildings; they can be owned by a variety of additional entities, from houses of worship, to a non-profit or private business.

The name energy democracy reflects the belief that current systems are energy monopolies, in which selected businesses have complete control of energy markets. Most definitions of energy democracy also address the need for economic transformation such as the creation of peer-to-peer marketplaces for electricity—a system in which one person sells electricity to their neighbors directly.[3] For example, if a farmer had a large solar or wind installation on their land, and sold power directly to nearby homes. All definitions set forth that not only does energy production need to transition to renewable sources, that a transition in the economics of energy systems is another paramount need, as an energy democracy approach recognizes "fossil fuel based energy systems and the associated massive corporate profits of large multinational energy companies have perpetuated inequities, exacerbated disparate vulnerabilities, and promoted widespread injustices among and within communities around the world,"[4] and that fundamental reform of this system is necessary to create more a sustainable and equitable system in this critical part of modern worlds. An energy democracy approach advocates for moving away from a model where one utility has the sole producer and distributor of electricity, and moving towards a model where individuals and communities are also energy producers, linking "the transformation/evolution of our energy system [to] other exciting aspects, and energy democracy through distributed generation and ownership is definitely at the top of the list."[5] Towards the end of this chapter, I discuss how Soulardarity took the first steps in the direction of community-owned energy production in Highland Park, putting energy democracy concepts into practice.

Researchers have compared energy democracy to the digital revolution, in which the growth of inexpensive cameras and computers combined with the growth of internet infrastructure, allowing everyone to become a media producer. Portals like YouTube, blogs, and podcasts started small and gained massive reach as well as respectability. This digital revolution changed more than the technology of news, radio, television, films, and media but also their economics and cultural dynamics. Everyone could become both a content producer and an audience member, and many could find new career pathways and become part of media ecologies. Many traditional media institutions adapted to this, and evolved with these changes; some did not. Similarly, a decentralized energy production system has the capacity to change the dynamics and economics of energy. Since decentralized energy is always renewable energy, energy democracy links the installation of solar, wind, geothermal and other renewable energy to equity, public participation, and democratic practices.

FROM ENERGY POVERTY TO ENERGY DEMOCRACY

A look at research on energy poverty and energy burdens, and how they are caused or supported by structural racism, will add more context to understand the unique history of Highland Park. The following statistics set into stark context how important the technological achievements of Highland Park's residents are, how lack of access to reliable and affordable energy intersects with equity and environmental justice issues, and how energy is affected by growth and loss of population.

Black Families Pay More for Their Light

Global research has developed many terms to describe specific aspects of how poverty relates to energy. Energy poverty is one of these, and means, in brief, when a family or household cannot access the energy they need. Some research ascribes energy poverty as primarily affecting the Global South, and uses other terms like "energy vulnerability" in the U.S., or "fuel poverty" in the United Kingdom,[6] all meaning essentially the same thing—not having enough power. Given circumstances in some U.S. cities, like Highland Park, any of these terms might be used accurately to describe a household that cannot meet its energy needs, regardless of the wealth of surrounding communities. Recent research recognizes the energy disparities in the U.S. and compares this inability to provide for one's own energy poverty to food insecurity, in which a family cannot afford to provide enough food to eat.[7]

For this chapter, I am using the term energy poverty to encompass all of these situations. Energy burdens are "dedicating inordinate amounts of income to energy services" which "can threaten a household's economic well-being over time, possibly by preventing a household from engaging in other economic activities or compounding existing economic hardship."[8] A 2020 study by Eva Lyubich and published by the Energy Institute at Haas (UC Berkeley) looked at energy expenditures in a sample of over 7 million families. It found a sizable race gap in energy spending, concluding that Black households have "higher residential energy expenditures than white households in the US," with Black households paying $200–$400 more per year for energy than white families.[9] This study found that the poorer the household, the bigger this gap. The disparity spans through all economic brackets for Black households, is consistent for homeowners as well as renters, and continues to be consistent even as income levels rise, only narrowing at the very highest income brackets. While the gap went down slightly, over the lengthy period this paper studied (2000– 2017), the reduction was not significant and the gap did not disappear.

A 2020 response to this study by Maximilian Auffhammer noted its importance in research on energy, as "so far, the energy economics community has done very little research on this topic." He emphasized the need for more research looking into *why* Black families in the U.S. pay more for their energy: do they use more? Do they live in areas with higher rates? The study speculated possible explanations, for example, if there was a similar gap in the energy audits that help lower costs, that could explain a structural imbalance that could be corrected. He also emphasized the need for ongoing research, and the need for that research to look at the *impact* of this kind of race gap—what does this mean for a family's long-term success and self-sufficiency? He also emphasized the need for research into how energy policies create this impact on low-income families, and correctly states "so long as this energy cost inequity persists, any future carbon/energy tax, or other policy that raises energy costs, is likely to increase energy expenditures more for Black households–especially poor ones–than for white households in the same income bin. And that's just wrong."[10]

Detroit Metropolitan Area: Energy (Dis)Service

The impact of energy burdens and higher costs for Black families is an ongoing challenge in the Detroit metropolitan area, which includes Highland Park. Highland Park residents are 91 percent African-American, with 46.5 percent of households living below the poverty line.[11] Their experiences with energy poverty and energy burdens illustrate how the previously mentioned studies affect people's lives on a day-to-day basis. The Detroit metro has the dubious distinction of having one of the highest national rates of power outages, with families regularly experiencing loss of power. A 2016 report by *The Eaton Blackout Tracker* listed Michigan as fourth in the nation for power outages, and the highest per capita level of power outages.[12] While this report was disputed by the local utility, many other industry reports gave DTE extremely low rankings. This includes a 2019 Electric Utility Performance Report by the Citizens Utility Board, a nonpartisan, non-profit advocating for residential customers of Michigan's energy utilities, which noted: "Michigan's poor performance on annual outage minutes per customer and middling or better performance on the number of outages per customer per year reflects that Michigan's power restoration following an outage is quite poor."

Lived experience corroborates this data, and shows the day-to-day negative consequences of power outages. Power outages are far more than an inconvenience—for a family just getting by, if an entire fridge full of groceries is spoiled because they lost power for two days, it can dramatically increase their monthly expenses. For people already living at or below the poverty line, this can be devastating. In some of the blackouts, DTE offered reimbursements for spoiled food, however the amounts were minimal, for example a reimbursement of $25 for an entire refrigerator of food that had spoiled.

Another noteworthy example of how a blackout can cripple a person's livelihood was in March 2020, during the very first days of the Covid-19 pandemic lockdowns. Detroit was one of the first hotspots for the Covid-19 pandemic, with Michigan joining New York and California, the sites for the other first major Covid-19 outbreaks, in implementing strict shelter-at-home rules. This necessitated turning to remote work and school. That same week, a massive wind storm hit southeastern Michigan, causing a record blackout that affected over a million people in DTE's service area, including parts of Highland Park. The power outage lasted for a week for some households.[13] In the early days of the pandemic, people were pushed overnight into becoming a one–hundred percent remote workforce, schools and colleges moved to online classes, and the entire nation turned to video calls to collectively try to understand what was happening. And Detroit and Highland Park lost power, cutting them off from information about the pandemic, protection, and the emerging public health initiatives, as well as from work, school, and connection with each other. This is precisely how the day-to-day reality of energy poverty and energy burdens work. People are impacted by inadequate utility service. Inconsistent power means a person is unable to go online, putting people at risk of losing their job or missing school. This means a greater risk of moving into energy poverty (if not already there) by pushing people down the ladder instead of helping them hang on. Electrical power at home is now critical for an individual to hold a job, attend school, and stay in touch with family—and losing power even for one day can be devastating for someone who might be struggling to keep their family (or themselves) afloat.

Echoing the research detailed above, more ongoing studies of energy burdens are important to deepen and expand the body of research on environmental racism, environmental justice, and how energy poverty intersects with how a family survives. As research grows on sub-categories of environmental justice, like food and justice, and access to water, studies must also further detail *solutions* to energy burdens and energy poverty. In this chapter, I hope to contribute to research on how people most affected by energy burdens can—and are—solving their energy problems using solar.

HIGHLAND PARK HISTORY: POPULATION GROWTH AND LOSS

What I'm trying to do here in Highland Park is because of my love for Highland park. I was born and raised here. I grew up here. My parents came here as one of the first few black families in Highland Park when it was affluent. I saw when the city was simply gorgeous. What I'm trying to do, I'm trying . . . in order to see the city come back . . . If I am successful, this is a win for the entire city.[14]

Highland Park, located within the Detroit city limits, has a unique, rich, and exhaustively researched history The short and relevant version is that in 1910, the Ford Motor Company opened the Highland Park Plant, which was the "birthplace of the moving assembly line" and "the first Ford plant to pay 5 dollar a day wage for an eight-hour workday."[15] This plant produced Model T cars in mass quantities, making them affordable to the average family. As the site for the first mass production of automobiles in the world, Highland Park had profound impacts on urban development, demographics, and economics for the entire region. First, it was home to the Ford Motor Company. Then, in the 1920s, Chrysler also established their global headquarters there. Geographically and figuratively, Highland Park was the center of Detroit, and fundamental to the Detroit metro area developing into a global economic driver. Highland Park's population swelled to over 50,000 in the 1930s and 1940s,[16] counting many executives and high-ranking employees of these companies among its residents and making it one of the wealthiest cities in the entire nation. Today, that original Model T plant, while still standing at the intersection of Woodward Avenue and Manchester Street, is vacant and has long been non-operational. Much of Highland Park's population, along with that of the Detroit metropolitan area, had to seek opportunities elsewhere, and now the city hovers at just over 10,000 residents.

What is most relevant from this history of automobiles to the solar-powered streetlights is the example of how a world-class industry grows from one innovation in design (the moving assembly line) and how applying that innovation in an appropriate way created more than jobs: it also built a regional economy, economic opportunities, and strong middle-class communities. Will innovation in solar do this as well?

The other relevant aspects are also the use of economies of scale to make new technology accessible. By building cars in mass quantities, the Ford Motor Company was able to sell them at a price many people could afford. The converse of this is true as well: population loss increases rates for services that rely on a shared economy, like a water system or electric grid. With millions of people paying into these systems when the metro Detroit population was large, they were affordable. But decades of closed plants and job losses took their toll, and the regional economy shrank. Families moved elsewhere. When people move away, so do city taxes. With a smaller tax base, the City of Highland Park had less money, and this was one reason they had, then in 2011, accumulated an overdue electric bill of $4 million. They negotiated a deal with DTE to lower their monthly cost of street lighting, and this deal resulted in the repossession of the street lights.

REPOSSESSION: HIGHLAND PARK LEFT IN THE DARK

One sunny afternoon in August 2011, a Highland Park resident, a historian and journalist, noticed that some of the streetlights in Highland Park had been clipped—they were just poles with no lights. He started taking photographs. Then, a few days later, he saw two contracted utility workers removing the streetlight that was in front of his home. He videotaped it. In a 7 minute video that he later shared with the media, he recorded these two workers unscrewing the metal pole of the streetlight from the base, loading it into a truck, and sealing the base with cement.

The basic facts, as reported in local media: in 2011, DTE Energy, the Detroit utility, repossessed over 1,100 street lights. The exact number varied, and some reports stated up to 1,400 light poles were taken. This was part of the negotiated settlement between the City of Highland Park and DTE to resolve the $4 million unpaid utility bill, which had accrued over many years. According to press reports at the time, the negotiated settlement reduced the monthly utility bill by $47,000, helping keep Highland Park out of further debt. But the damage it caused to residents who were abruptly left in the dark, with no warning or chance to alleviate the matter, was irreparable.

The Highland Park historian posted the videos to his YouTube channel. They were picked up by local news. His comments on the YouTube posting shed insight into the reaction to the lights being ripped out of the ground and captured:

> *Two contractors removed the DTE street light from in front of my home on Aug. 19, 2011. It took all of seven minutes to complete. The citywide removal of street lights was done under the euphemistic name "Highland Park Lighting Improvement Project," which was careful to downplay the fact that the lights on the STREETS, as opposed to SOME of those at the intersections, would NOT be replaced. DTE and the mayor have justified the removal on the basis of a debt-forgiveness and cost-saving arrangement. However, most residents are only learning about this program as the lights are being removed, and many are concerned about the possible ramifications on safety, particularly of our children and older residents.*[17]

The local news reports on the repossession of the streetlights shared residents' fears of an increase in crime. Three weeks later, the same local media reported a rash of break-ins attributed to the newly darkened streets.[18]

The story of Highland Park's repossessed streetlights has been kept alive for a decade, with stories running in national media outlets like New York Times and The Huffington Post, to public radio, regional media,[19,20]. It made its way into the pages of scholarly research.[21] Subsequent stories on how Highland Parker's responded by building solar-powered street lights have also attracted regional and national media attention.[22,23]

At the time, local residents went to the Highland Park City Council with their concerns about how the dark streets increased crime and diminished their safety on their blocks. In response, the Mayor and City Council recommended residents install lights in front of their homes to light their yards and streets. This suggestion effectively shifted the cost of street lighting from the city, which is supposed to be responsible for street lighting, to local residents, who had already paid for street lighting when they paid their taxes. In a city, where almost half of the families live below the poverty line, any increase in the power bill is too much of an increase.

INNOVATION: NEW STREETLIGHTS

Ten years after the street lights were repossessed, and a century after Henry Ford opened the Highland Park Plant, this three-mile square city is home again to an innovative new idea materializing into a tangible object. The nicknames reveal volumes: originally called the "City of Trees" because it was lushly forested until disease necessitated cutting down many trees. More recently, one writer dubbed Highland Park with the very unfair and rather dismissive nickname "Detroit of Detroit" due to its official status as Michigan's poorest city.[24] This nickname ignores the reasons behind Highland Park's decline, although it does have value in that it reveals something about the public perception of Highland Park, and whether issues like the streetlight repossession are seen in light of the larger forces at work undermining daily life for Highland Park residents. Comparatively, residents in the surrounding Detroit metro area have a median income of about $60,000; throughout the state, the median income is $57,000. They are well-off compared to Highland Parkers, whose median income is just over $18,000.[25] Today's Highland Park is due for a new nickname as residents resurrect the spirit of innovation, industry, and community.

Soulardarity, Ryter Cooperative Industries, and Parker Village

If you walk through Highland Park today, you will notice the unusual streetlights in the front yards of several homes and businesses. Their arching arms support small solar panels. Several have small signs leaning on them that say "Ask Me About Power," a slogan for Soulardarity, a Highland Park non-profit whose members created the solar-powered street lights and built community campaigns to educate the public about energy. Less than a mile from the empty Ford Model T plant, you will come to East Buena Vista and Brush Streets, where an especially eye-catching version of one of these solar-powered street lights rises from the parking lot of what used to be a school, complete with an illuminated plaque: "Parker Village."

Parker Village is well on its way to becoming Highland Park's second eco-village, following the example of nearby Avalon Village, another eco-village.[26] Parker Village is a planned residential community and co-working space founded by Juan Shannon. He describes it:

> *Parker Village will be Metro Detroit's first smart neighborhood . . . a beautiful, safe, secure, smart neighborhood. We just installed our first smart streetlight and the entire project is rooted in renewable energy, media, aquaculture and technology. A smart neighborhood is an area that utilizes a variety of technology. We're using solar, we're doing a blue infrastructure, which is the capture of rainwater and snowmelt in order to reutilize the water in other ways. We also are using geothermal, which we'll do the heating and cooling for our structures, and electric vehicle chargers. So it's a smart neighborhood . . . to break it down in a simple way . . . it is a neighborhood that is utilizing technology to use as little energy as possible and to re-use various products.*[27]

The first solar-powered building on the grounds of Parker Village was a small garden shed, about 10 feet long and 5 feet wide. On one side, a bright mural depicts the history of Highland Park, and features a polar bear, which was the mascot of the Highland Park High School before it closed.

As Juan showed me around the Parker Village grounds, he shared the story depicted in the mural:

> *So we got our fantastic mural here by a local artist named Waleed Johnson.*[28] *And it depicts some of the history of Highland park. You got the Ford plant and the Model T. We are still called "The City of Trees," the Highland Park High School mascot, the polar bear, and of course the Davidson Expressway. And then it shows the new growth and bridging from our history to the future of things that we're championing here at Parker Village with the aquaponics, the technology, with the renewable energy. And I love it. I think the mural came out beautiful and I just love it to death.*[29]

It takes a village to build a village. Soulardarity began its work in Highland Park through grassroots partnerships. It built working relationships with Ryter Cooperative Industries (RCI), a solar project management company and many other local partners, ultimately creating programs that led Highland Park to national recognition as a leader in practical clean energy innovation. Soulardarity's Deputy Director, Shimekia Nichols, shares how they got their start:

> *Soulardarity began when the streetlights in Highland park were repossessed in 2011 and it left the streets dark. Highland Park is in the heart of Detroit. You have to pass through it to get to downtown, to get over to the east side. It became a very unsafe place. The residents of Highland Park are very open to the*

ideas of going off the grid. When we explained to them what that looks like, they sometimes get a little caught up in the technical aspects, but they are definitely in support of the work. They've been in the dark for years now and they're glad to see that someone is advocating on their behalf.[30]

Jackson Koeppel was a co-founder of Soulardarity and its first Executive Director. In a 2017 interview with Michigan Public Radio, he described how the solar streetlights were designed to meet the immediate needs of residents: *"We wanted to demonstrate this technology because it was more affordable, it was more resilient, it would work during a blackout,"* he says.[31]

Soulardarity's first plan was large: they proposed to the Highland Park City Council that the city replace all 1,000plus repossessed streetlights with solar ones. However, this was an extremely expensive proposal for a city with many achingly important priorities, including ensuring water service. Soulardarity also began knocking on doors, holding regular meetings, and listening to community members, who had other ideas. Koeppel describes their switch in direction: *"A lot of members said listen, we're down, we want to see this large proposal go, but my street is dark, my alleyway is dark. I want to be able to actually do something right now,"* he says.[32]

The solar lights project is essentially community solar—a power production source, decentralized from a power company and also from the grid. As Michigan did not have state policies supporting community solar projects at the time, that was another hurdle. Koeppel continued: *"Everybody knows about [solar], everybody wants it, but nobody knows where to go to get it. And the costs of trying to get it individually are sometimes very high."*[33]

Founded by community members, Soulardarity had already carefully woven the concept of energy democracy into their practice. This meant fostering dialogues about energy rather than just installing technology, listening to what the community felt they wanted and really needed, and finding ways to decentralize, educate, and shift the economics of energy. They held energy fairs, barbecues, and regular community dialogues. They also worked with this community to see beyond the immediate energy burdens, and to understand the economics of energy, take part in decision making, and brainstorm ways to use solar to improve their lives in multiple ways.

As mentioned, Soulardarity's grassroots organizing began with meetings in Nandi's Knowledge Cafe, whose motto, "Knowledge is Power," is literally manifested through the solar streetlights. The floor-to-ceiling shelves are crammed with books on philosophy, health, music, and revolutions, the walls replete with paintings and wood carvings, and the entry crowded with racks of chic clothing and jewelry. Here, Soulardarity's founders planned public energy education events and door-knocking campaigns. They worked with RCI to address the immediate problem of the dark streets, eventually sourcing affordable solar-powered lights in a variety of wattages that could be easily installed over doors, porches, alleyways, and other areas that needed illumination. They started a program to sell small, individual-sized lights at bulk rates to community members after the City Council had instructed residents to install porch and yard lights, which effectively externalized the cost of street lighting from the city government to residents who paid taxes that are supposed to cover things like street lighting. Through solar lighting bulk purchases, they were able to lower the prices of the lights, employing the same principle of economy of scale that helped create the days of a wealthy Highland Park. The solar lights were easy to install and had powerful and efficient LED bulbs. Once a household bought and installed it, they could expect years of light at zero cost. This might seem to be a tiny step for solar integration, but it was major progress for community empowerment.

RCI was at the front line of Soulardarity's founding. They had previously been involved in other solar projects throughout Detroit, most notably by building an off-grid solar-powered charging station for D-Town Farms, a community garden supporting food security for African-American families. Ali Dirul, the founder and engineering director of RCI, shared his observations on the impact of introducing solar power: *"Although the [D-Town Farms] system was designed to deliver several thousand watts of power, we discovered that the idea was the most powerful thing. That the idea that an off-grid system could exist in the middle of this urban congested area created a whole new possibility for people that didn't exist before"* (Emphasis mine).[34]

Parker Village, Soulardarity's campaigns, and the other solar developments in Highland Park have attracted attention from regional and national media, many local news reports, Detroit cultural magazines, and statewide press. An energy think tank brought a group to its site to visit. The Detroit Innovation Fellowship lists Juan, Ali, Jackson, other RCI leaders, and the founder of Avalon Village—each using solar for community development projects—among their 2019-20 Fellows.

Collectively and individually, their work has extended far beyond the individual-sized solar-powered lights. RCI has expanded solar projects to solar-powered lawn mowers, solar-powered phone charging kiosks, and has grown its roster of private solar projects.

Five years after the street lights were repossessed, Solidarity presented a formal proposal to the Highland Park City Council to reinstall all of the street lights using solar-powered ones. Working with Parker Village, RCI, and many other partners Soulardarity has held two energy fairs—both standing-room-only events bringing together dozens of solar and renewable energy experts under one roof. In 2018, they held their first Gala celebration: coming together as a community to enjoy their gains. They have installed several solar-powered street lights, Soulardarity continues to receive press coverage in state and national media, and their staff present at an active roster of speaking engagements. Together they are all well-recognized leaders in energy innovation with an active and dedicated membership.

That is not all. The members of Soulardarity have a much broader vision, and continue to develop their plans. Shimekia Nichols, who became Soulardarity's Executive Director in mid-2021, sketched out a few of the possibilities:

> *I would definitely be satisfied if we could make Highland Park a model city for being 100% clean solar. We will be able to get Highland Park on board with 100% clean energy if we strengthen our educational programs. If we could get out in the community and do a lot more promotion, and if we have very strong collaborative relationships with other organizations and businesses.*

> *I think that [success] looks like having solar gardens, us having homes with interior lights or backlighting from solar, and a number of houses with solar panels. It will be great because it will lower the cost for the whole neighborhood. Adding the solar street lights is a huge step. And, just making a stance and setting the precedent that we won't be bullied by energy conglomerates.*[35]

BIG WIN: STATEWIDE COALITION FOR RENEWABLE POWER

In early 2020, the statewide partnership for improving renewable energy policy launched, of which Solidarity was a member. This campaign, titled, "Work for Me, DTE," reached out to community members to have them provide comments during a utility's public comment period. As a utility governed by extensive state and federal regulations, DTE is required to create a rate plan that lasts well over a decade, as mentioned at the start of this chapter. They submit it for public comment and state approval. The public comments were the crack in the otherwise impenetrable utility policy which the statewide coalition, including Soulardarity members, pulled into a huge fissure.

In a major victory, in late December 2019, the judge reviewed DTE's rate policy and listened to public testimony, and sent DTE's plan back to them. A front-page article in The Detroit Free Press ran a headline reading "Tiny Michigan nonprofit is taking on DTE—and it could have huge impact on the company." While this headline gave Soulardarity full praise, the reality is that the statewide coalition numbered close to 20 groups, including environmental and other community-based organizations throughout the state.

The Work for Me DTE! campaign generated thousands of public comments to the Michigan Public Service Commission and Legislature through the strength of the coalition, and also conducted legal and other types of advocacy. The result: a judge, in a nearly 200page opinion, told DTE that they needed to rewrite their submitted 15year rate policy, which had proposed rate hikes, and they should make it more efficient, and be far more precise about their goals for solar and wind power. Nine years after the streetlights were repossessed, Highland Parkers, using the premises of energy democracy, took a step to address energy poverty statewide, and to push forward policies supporting renewable energy for the entire state.

This is certainly not the end of the story: while it was a massive win, it is one opinion by one judge. The state regulatory body will still need to approve a final DTE plan, and DTE had over two years to make the revisions.

While it did not solve all of the energy problems facing Highland Parkers. But it does illustrate how this community put themselves more firmly in the driver's seat than they were on that August day a decade ago when the street lights were ripped out. Shimekia addressed how this changes the conversation on a community level, introducing the core idea of energy democracy to broader audiences: *"Energy democracy" speaks to a soulful concept—that a utility is not just a company selling products; it's a tool of society that should serve society . . . I believe, if the infrastructure is unreliable or archaic, the average person should not have to foot the bill for that.*[36]

As Shimekia steps up as Executive Director, she shares how their vision is highly personal: *"I do this work for my children and for my community,"* Nichols said. *"One day, I see Highland Park serving as a national model for overcoming adversity to become a healthy, growing, and thriving community powered by locally-owned energy."*[37]

In 2021, Soulardarity plans to install 10 more Wifi-enabled solar-powered streetlights and install solar panels on 30 homes. While the Highland Park City Council has not yet found the funding to replace all 1,000-plus repossessed streetlights, they have come a lot closer to embracing some of Soulardarity's plans. In 2021, a Highland Park Council member told the Energy News Network that the City Council now sees how Soulardarity's ideas are beneficial in many ways, and that their proposals support community-wide revitalization. Highland Park's challenges were due in large part to loss of population, and rebuilding infrastructure attracts new people to move to the neighborhood. As he put it: "People want to live there if they have this infrastructure that offsets energy costs."[38]

The Highland Park City Council did more than listen to Soulardarity's proposals, in Spring 2021, they discussed passing a "solar ordinance" to take the first steps in drafting guidelines for the inclusion of renewable energy in Highland Park. Additional suggested steps included battery storage, electric vehicle charging, and more solar-powered street lighting on larger projects.

As Jackson steps down from his decade-long tenure, Soulardarity is seeing solid gains. Interviewed about these gains as he handed off the office to Shimekia, he discussed the heart of their progress:

> *Soulardarity offers solutions to these kinds of problems and seeks input from Highland Parkers instead of just making noise or critiquing City Hall, [We] knock on most doors in Highland Park every year for one reason or another. We have done deep work to develop demands that reflect the broad community. For every ounce of pressure we have put on the city, we have brought answers and solutions. We're here to stay. We're dedicated.*[39]

Each gain represents another new challenge, and in their annual report, Soulardarity reported that they had "filed incorporation paperwork for Polar Bear Sustainable Energy, a member-owner cooperative that will provide holistic energy services in Highland Park and beyond."[40] While this is an initial step towards building a community solar project in Highland Park, their solid track record speaks to how their mix of door-knocking, technical education, working with the community, and taking small but mighty steps closer is a strategy that is already proven to work. In a few years, Polar Bear Sustainable Energy might just be a living, energy-producing reality that brings the promises of energy democracy principles to Highland Park.

CONCLUSION: KNOWLEDGE IS POWER

At the time of writing this, Soulardarity continues building its programs and membership, and expanding its reputation and resources. RCI continues building its solar consulting business and products, as well as its reputation for solar expertise in metro Detroit. They saw a surge of business during the Covid-19 pandemic shutdown, particularly for solar-powered Wifi products.[41] Parker Village is becoming the eco-village that Juan envisioned. Most recently he started building the solar-powered cafe and conducted hands-on training for solar installers on his site. The Cafe at Parker Village has a constructed framework with the kitchen equipment due next. RCI designed the off-grid solar power for the cafe, a 4.5 kilowatt system that will be mounted on the roof of a deck, to create a covered area for the cafe.

The collective efforts of everyone addressed in this chapter, and many whose work was not possible to discuss, highlight how applying the ideas of collective economy, public participation, and democracy can address enduring issues that seem permanent, like energy poverty. They show that these issues are within our power to change. Their work also shows how underlying institutional issues create problems related to socioeconomic conditions, like energy poverty and energy burdens, and a lack of understanding of these bigger causes holds back development and new growth.

Another important issue revealed is how attitude and perception about Highland Park affect it, as Juan eloquently stated:

I will say that sometimes when you hear about a place or read about a place, you only have a certain amount of time to tell the story. And so people get the impressions of places that they've never been to. And that's their only impression of that place. Highland Park, just like Detroit, has this reputation like . . . it's just so dangerous. And it can't be further from the truth. There's danger everywhere. Highland Park, for all of its flaws and for all of the blight that has been caused by various reasons . . . people just think it's really dangerous. Leftover, a forgotten place. And there are still near 10,000 people here and a lot more around the world that love this place. This 3mile little area was the most affluent city in the United States. When you have 2 corporations leave the city, taking about 70% of the tax base with them, any city would start having problems with that. That's a lot of money."[42]

Parker Village's street light stands out in modern contrast, its sharp edges shining in the parking lot. The LED sign glows when the sun sets, and the street light is dim until a car approaches, and then its motion detector turns it on to fullest brightness. When asked how his neighborhood has reacted to his solar-powered streetlight, Juan said:

Oh my goodness. I've had people come up and try to hug me. They are very happy that there's some light over here. I've got neighbors on the other side that are a little jealous and upset because we haven't put one over there yet. But I got a surprise for 'em. We're gonna get some up . . . prayerfully in the next few months and just continue to work on the project.[43]

Parker Village—the result of Juan's vision and entrepreneurial sensibilities, Soulardarity's solid programs, and RCI's management and knowledge—attracts attention to a less-common use of solar. Solar-powered lighting creates safety, costs nothing after installation, and works despite power outages. While all the issues of energy poverty are not solved by street lighting, a little bit of solar seems to go a long way. During the Covid-19 pandemic, Juan made Parker Village a food distribution center for Covid-19 relief. He continues its schedule of regular volunteer building days, fairs, and other events. He has a long way to go before fully opening the doors, however, as mentioned above, he's also here to stay and dedicated: When asked when he would feel like Parker Village was a success he said:

I will feel like we've at least hit a complete job when we have a bunch of graduates from our STEAM courses, from our urban gardening, from our aquaponics courses. And those people go on to do good things in the world because they had a chance to be at Parker Village. I will feel that, you know, my job is complete and I'll be on to the next creative process.[44]

If you are in Detroit, take a trip to Highland Park, look for the solar street lights and check out the growing site of Parker Village. Perhaps you can stop by Nandi's Knowledge Cafe for lunch and a moment of reflection on how knowledge, power, and talking to your neighbors can lead to a long-lasting change.

ACKNOWLEDGMENTS

The Work for Me DTE! Campaign partners include: East Michigan Environmental Action Council, Engage Michigan, Ecology Center, Meta Peace Team / MCHR, Good Jobs Now, MI Citizens For Conservation, Coalition To Oppose US Ecology, WACO, D2SOLAR, Sierra Club, MI Interfaith Power & Light, Good Jobs Now, Soulardarity, Michigan Environmental Justice Coalition, Citizen's Resistance Against Fermi Two, We Want Green Too!, We The People Detroit, and Great Lakes Bioneers Detroit.

NOTES

1. Soulardarity website. "Why Energy Democracy." www.Soulardarity.com. Accessed June 1, 2021. https://www.soulardarity.com/why_energy_democracy

2. Hasberg, Kirsten S., and Tegilus, Bob. "Energy Democracy Defined," *This Week in Energy Podcast.* August 9, 2013. https://archive.org/details/hpr1310. See also: https://vbn.aau.dk/en/publications/energy-democracy-defined-episode-of-this-week-in-energy-podcast.

3. Hasberg and Tegilus, 2013.

4. Stephens, Jennie C. "Energy Democracy: Redistributing Power to the People Through Renewable Transformation." *Environment: Science and Policy for Sustainable Development.* 61, no. 2 (2019), 4–13. doi: 10.1080/00139157.2019.1564212

5. Gerke, Thomas. "Energy Democracy — Video & Campaign." *CleanTechnica Campaign Paper.* May 14, 2013. https://cleantechnica.com/2013/05/14/energy-democracy-video-campaign

6. Sonal, Jessel; Sawyer, Samantha, and Hernández, Diana. "Energy, Poverty, and Health in Climate Change: A Comprehensive Review of an Emerging

Literature." *Frontiers in Public Health.* 7 (December 12, 2019), 357. doi: 10.3389/fpubh.2019.00357

7. Bednar, Dominic J., and Tony G. Reames. "Recognition of and Response to Energy Poverty in the United States." *Nature Energy* (2020): 1–8.

8. Bohr, Jeremiah, and Anna C. McCreery. "Do Energy Burdens Contribute to Economic Poverty in the United States? A Panel Analysis." *Social Forces* 99, no. 1 (2020): 155–177. doi: 10.1093/sf/soz131

9. Lyubich, Eva. "The Race Gap in Residential Energy Expenditures." 2020. https://haas.berkeley.edu/wp-content/uploads/WP306.pdf

10. Auffhammer, M. "Do Black Households in the U.S. Pay More for Their Energy?" *Energy Post.* Last modified July 23, 2020. https://energypost.eu/do-black-households-in-the-u-s-pay-more-for-their-energy/

11. U.S. Census. *Highland Park, Michigan, 2019 data.* https://censusreporter.org/profiles/16000US2638180-highland-park-mi/

12. French, Ron. 2017. "Study Claims Michigan Has More Blackouts Than Most States. Not True, Says DTE." *Bridgemi.Com.* https://www.bridgemi.com/michigan-government/study-claims-michigan-has-more-blackouts-most-states-not-true-says-dte.

13. Fournier, Nicquel Terry, and Holly. 2015. "Outages Drag Past a Week for Some DTE Customers." *The Detroit News.* March 15, 2015. https://www.detroitnews.com/story/news/local/oakland-county/2017/03/15/detroit-edison-power-outage-stragglers/99230388/.

14. Shannon, Juan. "How to Power a City." Interview by Melanie La Rosa. May 2018.

15. Detroit Historical Society. 2021. "Ford Highland Park Plant | Detroit Historical Society." *Detroithistorical.Org.* https://detroithistorical.org/learn/encyclopedia-of-detroit/ford-highland-park-plant.

16. U.S. Census.

17. Lee, Paul. "Highland P Streetlight Removal." *Paul Lee YouTube Channel,* 2011.

18. Molony, Ray. "Major Study Finds Outdoor Lighting Cut Crime by 39%." *LEDs Magazine,* Originally published on *Luxreview.com,* March 1, 2018. Republished through partnership, April 20, 2020 https://www.ledsmagazine.com/architectural-lighting/article/16701511/major-study-finds-outdoor-lighting-cut-crime-by-39

19. Costantini, Fabrizio. "Cities' Cost Cuttings Leave Residents in the Dark." *New York Times,* Dec. 29, 2011. https://www.nytimes.com/2011/12/30/us/cities-cost-cuttings-leave-residents-in-the-dark.html

20. Eichler, Alexander. "Highland Park, Michigan Tearing Out Its Streetlights to Cut Costs." *Huffpost.Com.* 2012. https://www.huffpost.com/entry/highland-park-sreetlights_n_1079909.

21. Brennan, Shane. "Visionary Infrastructure: Community Solar Streetlights in Highland Park." *Journal of Visual Culture* 16, no. 2 (2017): 167–189. doi: 10.1177/1470412916685743

22. Daily Detroit Staff. "This City Had Their Streetlights Repossessed. Now Some Are Getting Light Thanks To Solar Power." *Daily Detroit.* 2017. http://www

.dailydetroit.com/2017/04/25/city-streetlights-repossessed-now-getting-light-thanks-solar-power/.

23. Cwiek, Sarah, and Steve Carmody. "Highland Park Residents Lighting Up Their Streets With Solar Power." *Michiganradio.Org*. 2017. https://www.michiganradio.org/post/highland-park-residents-lighting-their-streets-solar-power.

24. Binelli, Mark. *Detroit City Is the Place to Be: The Afterlife of an American Metropolis*. Macmillan, 2012.

25. U.S. Census. *Highland Park, Michigan, 2019 data*. https://censusreporter.org/profiles/16000US2638180-highland-park-mi/

26. *The Avalon Village Official* website. http://theavalonvillage.org/, accessed December 23, 2020.

27. Shannon, Juan. "How to Power a City." Interview by Melanie La Rosa. November, 2018.

28. Art-Ops. "Community ART Highland Park." *Art-Ops.org*, accessed December 14, 2021. https://art-ops.org/community-arts-highland-park

29. Shannon, Juan. "How to Power a City." Interview by Melanie La Rosa. November, 2018.

30. Nichols, Shimekia. "How to Power a City." Interview by Melanie La Rosa. September, 2017.

31. Cwiek, Sarah. "Highland Park Residents Lighting up Their Streets with Solar Power." *Michigan Public Radio*, The Environment Report. February 21, 2017. https://www.michiganradio.org/post/highland-park-residents-lighting-their-streets-solar-power

32. Cwiek, Sarah, 2017.

33. Cwiek, Sarah, 2017.

34. Dirul, Ali. "Inspiring Possibilities Through the Power of Solar." *TEDxDetroit*, TEDx Youtube Channel. October 17, 2018. https://www.youtube.com/watch?v=jcqt9MgkWbA

35. Nichols, Shimekia. Interview, September, 2017.

36. Laitner, Bill. "Tiny Michigan Nonprofit is Taking on DTE — and It Could Have a Huge Impact on the Company." *Detroit Free Press*. January 16, 2020.
https://www.freep.com/story/news/local/michigan/wayne/2020/01/16/dte-soulardarity-energy-democracy/4429251002/

37. Soulardarity Press Release. "Michigan Nonprofit Soulardarity Names New Executive Director." *AlternativeEnergyMag.com* and *Soulardarity.com*. April 22. 2021

38. Perkins, Tom. "Michigan's Energy Activists Start to See a Decade of Work Pay Off in Highland Park." *Energy News Network*. May 4, 2021. https://energynews.us/2021/05/04/michigans-energy-activists-start-to-see-a-decade-of-work-pay-off-in-highland-park/

39. Perkins, Tom. "Michigan's Energy Activists Start to See a Decade of Work Pay Off in Highland Park." *Energy News Network*. May 4, 2021. https://energynews.us/2021/05/04/michigans-energy-activists-start-to-see-a-decade-of-work-pay-off-in-highland-park/

40. Soulardarity.com. "2019 Annual Report." *Soulardarity.com*. Accessed June 15, 2021.

41. Roff, Kate. "COVID-19 Shines a Light on Accessible Solar Power's Role in Detroit Communities." *Model-D Media.* December 15, 2020.

42. Shannon, Juan. "How to Power a City." Interview by Melanie La Rosa. November, 2018.

43. Shannon, Juan. "How to Power a City." Interview by Melanie La Rosa. November, 2018.

44. Shannon, Juan. "How to Power a City." Interview by Melanie La Rosa. November, 2018.

Chapter 10

Las Vegas, NV
100% Renewable Energy for City Power

INTRODUCTION

Las Vegas is one of America's most unique cities. Located in a wide valley, it rises out of the desert landscape. From this vantage point, the idea of the "built environment" is evidenced perfectly in the collection of hotels, casinos, the Ferris wheel, the mock Eiffel Tower, and all the other uncommon shapes of the buildings, all blending into the unmistakable Vegas skyline. Suburbs spread out from this city center, although soon enough, the desert takes over. The metropolitan area appears as a shining oasis of activity from the city's outskirts, surrounded by the sweeping valley floor, red rock canyons, and surrounding mountain ranges. The city, known for nightlife, glitzy shows, and shopping, is also in the center of a large region with spectacular outdoor activities. Las Vegas is only a 2hour drive from the North Rim of the Grand Canyon, and the area appeals to tourists for its slot canyons as well as for its slot machines.

At the time of writing, hundreds of cities and states across the United States have declared renewable energy goals ranging from goals of 50 percent, 80 percent, and, for many, up to 100 percent renewable energy for their city or state. But, as cities across the nation establish these laudable and ambitious goals, what does it entail to reach these? What sort of leadership, business partnerships, and investment?

Given the discussion in previous chapters about the challenges that utilities faced when they first started adding distributed generation sources to their grids, such as managing a swell of solar at the middle of the day, what insights can Nevada's experience yield for other places?

CHAPTER SUMMARY

During research and production, in 2017 and 2018, many cities and states announced commitments to add significant amounts of renewable energy to their grid. However, Las Vegas was ahead of this trend and hit an important milestone of using 100 percent renewable energy for city services in December 2016. While a few other smaller cities had already hit the 100 percent renewable energy mark, Las Vegas became the first large city in the country to use 100 percent percent renewable energy for all of its city services: streetlights, police and fire departments, city buildings, schools, and all other city-owned buildings. With this nationwide trend on the upswing, this chapter discusses how Las Vegas achieved a goal that is on its way to being universal for cities and states. It also discusses the physical risks that Las Vegas faces related to electricity and being a desert city. Among these are dangerous and prolonged high temperatures, which drive people to inside air-conditioned buildings. All that air conditioning creates stress on the electrical supply. Air conditioning in Las Vegas is more than just a luxury, as dangerously high temperatures can severely impact many vulnerable people, including the elderly and young children.[1]

The same intense sun creates dangerous temperatures which is also one reason why Las Vegas was able to transition. With about 300 days of sun each year, Las Vegas is not a surprising candidate for the first large city to hit 100 percent renewable energy. But, was it only their amount of sun that made this possible? Las Vegas ties with Phoenix for the third sunniest city in the nation,[2] and other cities in the top 20 sunniest, like Los Angeles, Albuquerque, Sacramento, and even El Paso, also seem like candidates to reach the 100 percent renewable energy mark. So, what did Las Vegas do that got them to this goal, which so many cities and states aspire to reach? This chapter looks at how Las Vegas collaborated with the utility, in light of tremendous population growth and the intense heat, to keep families, homes, schools, offices, millions upon millions of tourists, the neon buildings, and nightlife consistently and cleanly powered.

METHODOLOGY

I contacted the City of Las Vegas after reading news reports about them hitting the 100 percent renewable energy mark. I contacted the City Manager's office, who put me in touch with the Mayor's press director and the city planner who managed and implemented the Las Vegas sustainability initiatives. In 2016, it was still a relatively common belief that running on 100 percent solar power was difficult technically. My questions for the project manager were technical, while my questions for the Mayor were about the policy and background. I also contacted NV Energy, based on their strong collaboration with the city, and their communications director responded positively. He arranged for me to visit NV Energy's solar sites at Nellis Air Force Base and Higgins Generating Station and interview NV Energy's V.P. of Sustainability and an engineer at the Higgins Generating Station.

The interviews with Mayor Goodman and NV Energy were recorded in September 2017. I also researched their projects and the public information they shared on the City of Las Vegas website about their solar energy use and many other sustainability measures, such as ways to save water and reduce auto emissions.[3]

The entire filming process was three days in September 2017. The interview with Mayor Goodman took place at City Hall, and the additional interviews were on the NV Energy sites. Since the city had already hit the 100 percent renewable energy milestone, while many other community-based and industry-led solar and clean energy initiatives were taking place, transitioning city services had already concluded. I continued to monitor progress in Las Vegas online for changes.

As with any project with an expansive and diverse topic, there were many limitations to my research. This is by no means a comprehensive account of every way solar is being used in Nevada. Tribal governments have installed solar farms, Nevada has one of the highest rates of residential solar use in the country, and there are many solar and green energy activist organizations. In addition, casinos are installing some of the nation's largest rooftop solar arrays. My research focused specifically on the city itself because I was following the goals that many cities were setting for 100 percent renewable energy. Because this is such a narrow focus in a state with overall progress, I have included brief descriptions of broader statewide solar and energy initiatives at the end of this chapter, including tribal use of solar, military use of solar, and nuclear waste storage. These provide readers a glimpse of the surrounding context for the City of Las Vegas.

CITIES, GOALS, AND PERCENTAGES

Cities and states are setting goals for 100 percent renewable energy, or other large percentages, at unprecedented numbers. In 2018, researchers at Columbia University's Earth Institute published a study looking at the number of cities that have done this. This study covers 56 cities that made 100 percent renewable energy commitments in a campaign launched by the Sierra Club; in 2017, this campaign had a record year, with 30 cities signing up, compared to 5 in 2016 and even fewer in 2018.[4] Of course, that does not mean that only 56 cities have made these commitments—the Sierra Club's campaign is one of many such initiatives. In fact, Las Vegas's efforts began in 2008, 3 years before Mayor Carolyn Goodman took office when the previous mayor signed onto a similar climate proposal through the U.S. Conference of Mayors[5] (see next section).

There are quite a few proposals, campaigns, and initiatives that cities or states can turn to in making their energy commitments. This report by Columbia's Earth Institute is just one starting point providing a snapshot to understand the broader view of how various cities are planning and implementing energy transformations. It also adds context to renewable energy commitments nationwide and concludes that while there were a variety of motivations for those 56 local governments, they all agreed on one thing: the opportunity to link socio-economic development to clean energy and ensure that both goals are part of this imperative transformation.

What is a good baseline number from which to measure 100 percent ? What is the national percentage for use of clean energy? There are entire industries that measure and track the amount of clean energy that cities and states use: these include federal energy authorities like the Energy Information Administration (EIA) and industry organizations like the Solar Energy Industries Association (SEIA) and the American Wind Energy Association (AWEA). In researching for the film and the book, there were times that figures from various bodies were inconsistent with each other. They also employed a variety of metrics to count energy production. For example, some statistics measure the total number of solar panels installed for the state (called "installed capacity"). Others measure the electricity produced by those panels and look at whether the installation reached its capacity. If there were many cloudy days and storms, the power production might be lower than the capacity of the installed panels. Some reports focused on the amount of rooftop solar installed on residences, while others measured the solar farms owned by utilities but not solar on private homes. Wind energy was divided into onshore production, which was significant, and measured in many of the same categories as solar: installed capacity, proportion of total demand, percentage of power, and actual production versus installed capacity. Measuring the tiny amount of offshore wind in the U.S. is usually discussed in terms of the *potential*, which are estimates of power production; going one step further than estimates, some wind power metrics measure the amount of solicitations issued by states.

These numbers can be confusing. As a case in point: at the time of writing, Hawaii is reported as having the highest *proportion* of solar power, with 33 of Hawaiians getting their electricity from rooftop solar in 2019.[6] However, another source, from 2020, lists Hawaii at 14 percent, and California with the highest *percentage* of electricity from solar, at 22 percenti.[7] While these amounts differ quite a bit, they help in establishing a starting point that the most significant solar states hover at around one–fifth to one–third of their electricity from the sun.

These numbers speak volumes in terms of solar progress when compared to the national amount of electricity coming from solar, which was 3 percent in 2019. These metrics provide a clearer picture of how far there is to go. Most importantly, they show even more clearly how important it is when Las Vegas (or any city or entity) actually hits 100 percent renewable energy for their electrical use.

Nevada consistently ranks high on all the "best in solar" lists, particularly rankings for the *proportion* of the state's electricity from solar. Nevada is often ranked number 1 for proportion and per capita use of solar. This is because it has a relatively small population and an extensive amount of solar through many sectors—utilities, private households, private businesses, and tribal governments. The same 2020 report that listed California as number 1 for the highest percentage of solar named Nevada as number 3, with Massachusetts coming in second.[8] In fact, Nevada's state legislature passed a bill in 2019 committing to 50 percent of energy from renewables by 2030. It is well on its way: currently, the state is at about 20 percent renewables, with 50 percent of that being from solar, a little bit over 40 percent from geothermal, and a small amount of hydropower from the Hoover Dam.[9]

Back to that baseline number. If 22 percent is the highest proportion in the country for solar (keeping in mind that the 100 percent renewable energy goals include *all* types of renewable energy including wind, hydro, and geothermal power), it might seem as though even early-adopter states still have quite a distance to go. And they do—however, have all come a long way, and there are far more resources now than ever before to support solar and clean energy integration. The entire industry has come very far in the past decade—the national percentage of solar-generated by utilities in 2019, reported by the U.S. EIA, was 1.7 percent; however, that figure does not include rooftop solar. The SEIA reported that in 2019, solar was 3 percent of all U.S. electrical generation. More telling is comparing this to the SEIA's 2010 number: a decade ago, solar was 0.1 percent of all U.S. electrical production.[10]

As of 2019, 11 states plus Puerto Rico and the District of Columbia have officially set 100 percent clean energy goals, including Nevada. Hundreds of other cities and counties have as well. These are not unrealistic fantasies: as of 2019, the UCLA Luskin Center for Innovation issued a detailed report listing 72 U.S. cities and counties which have achieved 100 percent renewable energy.[11] Much research is emerging to track these achievements, and it is an exciting development with significant potential to expand to other cities, states, and counties.

BACKGROUND

Las Vegas did not go 100 percent renewable energy for all of its electricity—it was just for city services. This includes schools, fire departments, the many city buildings, police stations, street lights, and parks. The image that most people have of Las Vegas is the Strip, with all of the casinos and flashing neon. As the scope of the film focused on the city itself, it was not possible to include casinos, private businesses, or the extremely healthy market for residential solar. Many of the Las Vegas casinos have actually installed massive rooftop solar systems, and in fact, at the time of filming a rooftop solar system on the Mandalay Bay casino, at 8.3 megawatts, was considered the largest in the United States (although that has probably changed by the time you are reading this book).[12]

The other note is that Las Vegas is considered the first *large* city to be 100% renewable energy for its city services. Other cities also claim that distinction, such as Aspen, CO and Burlington, VT. They are both far smaller—Aspen's population is 7,300, and Burlington's is about 43,000—and do not have nearly the population challenges that Las Vegas has. The very first city to hit the 100 percent renewable energy milestone was Greensburg, Kansas.[13] In 2007, a tornado ripped through this town of 1,500 people and destroyed it. The town was entirely rebuilt and transitioned to using wind energy. As cities of all sizes set these renewable energy goals, it is essential to look at many different models of how other cities achieved 100 percent renewable energy.

LAS VEGAS—GAME PLAN

To say the sun in Las Vegas is intense is an understatement. Nevada's sun is not a warm glow shining softly on you. Rather, it is a forceful blaze that presses on you. It is relentless. In the middle of the day, the glare makes it difficult to see. Add to this, on windy days, the desert gusts blow grit into your eyes, and it is easy to understand the allure of being inside, particularly in an air-conditioned office, school, home, or resort. The sun's palpable nature and intense heat are also a constant reminder that solar energy is a never-ending, free-flowing resource with a very predictable and measurable schedule.

The City of Las Vegas serves a population much more extensive than its official residency. In 2016, when they hit the 100 percent renewable energy mark, the population was approximately 600,000 for the City of Las Vegas itself. But there were over 2 million residents in the metropolitan area, including Clark County. The Strip, where the largest casinos are located, is actually in an unincorporated township called Paradise. And the official population number is tiny compared with the 40 million-plus tourists that visit Las Vegas each year, a constant draw on its resources.

In an arid desert, temperatures regularly rise above 100 degrees, and a city requires massive amounts of water and power for heating, cooling, and lighting. Las Vegas has been growing for decades: from 1990 to 2000, its population grew by an astronomical 85 percent. In the next decade, this continued by another 22 percent, then 11 percent more by 2019. As discussed in other chapters of this book, particularly chapter 9 about Highland Park and population loss, there are close links between population and energy. This fast-growing population created another motivation for powers-that-be, who needed to find energy sources to meet this demand and wanted to do so without creating more pollution.

Demand and the Duck Curve

Tourism is one of the most important considerations regarding energy. Population creates what the industry calls "demand," which is the aggregated energy use of everyone and everything drawing from the grid. Demand is more than just use; it is the need to predict the use of an entire city: when people come home from work, demand rises because they all turn on their air conditioners, televisions, and microwaves at roughly the same time. In addition, tourists flock to cities, coming for round-the-clock activities and creating a retail load beyond what the city residents' use. In a tourist capital like Las Vegas, that demand is constant, and utilities know the patterns of demand. However, a consistent rise in the actual population means demand will also rise consistently. As a result, the utility will need to produce more electricity to meet the demand for things that tourists do not use, like schools, residences, offices, and daily living.

Predicting and meeting demand is how a utility ensures there are no brown-outs, black-outs, or losses of power. People like to come home, flip on the A.C. and the T.V., and relax. The job of a utility is to ensure they can do that without knocking out the power on their block or for the city. This predictable curve has come to be referred to in the industry as the "duck curve"[14] because it sort of resembles a duck. One of the major issues with solar power generation was that the electricity was produced during the day. Even in a state with very predictable sun, like Nevada, when the sun starts setting, the power production from solar also goes down. Unfortunately, demand rises right around the same time—and all those air conditioners go on as people arrive home just before sunset. As discussed in previous chapters, as energy storage systems come into more widespread use, this will likely not be the kind of technical issue that it was, and utilities are already learning to manage this swell and ebb of power production and demand.

The Fabulous Mayor Carolyn Goodman

Mayor Carolyn Goodman led the initiative in which the city and utility collaborated for consistent and clean power. During an interview, she shares how and why it began. Mayor Goodman mentions her husband, Oscar Goodman, who was Mayor of Las Vegas for 12 years before she was elected. Arriving at Las Vegas City Hall, a visitor can see the values of sustainability reflected in its design. An energy-efficient building unveiled in 2012, graceful solar "trees" line the front of City Hall. They are functional solar generators with a native plant and cactus garden surrounding them. There is also a solar array on the roof. The building's design is reported to save $400,000 in energy costs.[15]

Upon arriving at the Mayor's office, a huge sign outside of her otherwise understated office states "Welcome to the Fabulous Mayor Goodman's Office." She says it is a holdover from her husband, and luckily they didn't have to change the name.

> When my husband and I came to Las Vegas from the East coast in 1964, there was less than a hundred thousand people. Now we're over 2.2 million,[16] and that has nothing to do with the 43 million visitors we get every year, who come because of our conventions, because of our wonderful climate, for the wonderful entertainment that we offer here.

When asked about her motivations to lead this initiative, she spoke to how broader sustainability issues, such as eliminating toxic waste, were part of the reason:

> It is all about how people are going to get here, stay here, live here, and be able to visit here and not create more pollution and filth, but make sure it's a clean way of living. We're looking into everything that matters today and for the future. . . . What can we learn from the past and problems and nuclear issues, and our own testing of nuclear waste products? I think it's all about the different aspects that make up a healthy climate and a sustainable earth.

Mayor Goodman also spoke about the pivotal role national agreements played in starting this initiative and how her husband took the first steps:

> In 2008, my husband signed onto the U.S. Conference of Mayors Climate Protection Proposal.[17] And we made a total commitment because we saw what was happening in the world. We knew about pollution. We knew about issues with recycling and water conservation. And so we've been gung ho ever since.

Given this fragile environment and fast-growing population, Las Vegas has implemented many sustainability measures in addition to solar power. These broader efforts include clear choices such as far-reaching water conservation efforts, as the Mayor describes, as well as less visible approaches, like diverting trash from landfills, building city parks irrigated with drip systems, building a monorail to cut down on traffic, and installing LED lights to reduce the electrical use.[18] Their efforts have won Las Vegas several sustainability awards, including the 2014 Mayors' Climate Protection Awards from the U.S. Conference of Mayors. How a city integrates these various sustainability measures, particularly how they mesh with solar and other types of clean energy, are excellent topics for future research.

Moving forward on solar, Las Vegas began in 2008, well before the nation was turning to 100 percent renewable energy goals. I asked how Las Vegas got past the political loggerheads that plague most discussions of renewable energy. The Mayor's response:

Well, I think at the city government level there has been very little political polarization. We've had a little bit in the past couple of years, but I think everybody has been very much made aware . . . [that] this is a nonpartisan issue, and we really need everybody cooperating. We did hire individuals, who already were in the educational arena learning as much as they could about conservation, about renewable energy, about water, about recycling[19] . . . it really was spreading the word. It was getting all parties, private and public, together on it.

She elaborated on the commitment to making this a nonpartisan issue:

The Office of the Mayor and City Council is supposed to be nonpartisan, and we pride ourselves in that. And, as a matter of fact, I am a registered nonpartisan. Not an independent, that's a party, but a nonpartisan. Because I really believe that this body, this organizational government body, is here to take care of all the people—regardless of race, religion, culture, and political party. It's for everybody. And so we've been operating that way. Even though I know my council has a political bent one way or another, when they come in here to work for the city, I remind them this is what's good for the whole.

Goals are easy to set but can be very hard to achieve. Sometimes taking the first few steps is the most challenging part, particularly in city governments where bureaucracy, politics, and other priorities can intervene. When asked how they began, the mayor responded:

You cannot just go headlong into something. You need a team of people that believe the same thing to pull them together. And then to figure out both a short-term plan, and a long-term plan . . . you just don't go rushing into something like "I need this, I want it, I'm gonna do it, and who cares about anybody else?"

> So, it's a process. The first thing you have to realize is, I believe we have a problem." And find out what that area is. Is it something that's fixable? And then how do we get private partners involved in this, along with our public and our government? Not only our city municipal government but our state. And of course, whenever it can be, our federal government.

For energy, public-private partnerships are a standard business model, and I asked her to elaborate on how that worked:

> We've had a wonderful time with our energy partner, NV Energy, who's supplied the lion's share of the solar panels that are just outside the city limits here. That made us a hundred percent renewable energy, we are the largest city in the entire country, if not in the entire world that has all of its city buildings, the parks, all of our lights—we have 52,000 LED lights—and this has made it possible for us to be a hundred percent renewable energy.[20] We completed it . . . December 31st, 2016, a couple of months ahead of schedule.

The City of Las Vegas' website shares the details of their solar transition and a clear sense of pride in this achievement, which not all cities share. I asked what the public's reaction had been:

> The media was just thrilled to death. A lot of people of course around the country said: "What do you mean by that? Do you mean your whole city is renewable energy?" All we can handle is that of which we're in charge, which are our buildings, our parks, our lighting, and areas like that. We can't do this in the private sector . . . we have to entice the private sector to participate for themselves. Then you become city-and county-wide, and that makes that difference.

I asked her to speak to any response from city residents and whether they contacted the mayor's office to share their reactions:

> Well, today, you get tweets, you get letters, you get emails—hundreds of emails and phone calls. The city was so proud. You see, we're different here in Las Vegas . . . Here, everything operates almost entirely all the time. And there's a great deal of pride in choosing to live here . . . So the people who have moved here or choose to move here, which we did . . . was because we wanted that good feeling of doing something to make a difference.

Technology and Collaboration

The steps Las Vegas took were first to plan the project and secure funding through the Recovery Act of 2009 and other state and local funding sources.

With that funding secured, the City of Las Vegas worked out an agreement with NV Energy in 2015 to buy 100 percent of its "retail load" (the amount of electricity purchased from the utility) from solar facilities, including a new one outside of Boulder City, NV. When a city buys power, it covers everything from the streetlights and City Hall to office buildings, schools, fire departments, police stations, parks, and all public facilities—which is a lot of electricity.

This actual switch to solar is a concise section in this chapter, as it involves a business agreement. The City of Las Vegas signed a purchase power agreement to buy solar energy. Because their retail load is very large, it meant that NV Energy could authorize the building of more solar power plants. At the time, NV Energy had just opened the new Boulder City solar power plant. When they did, they also closed a notoriously polluting, coal-burning power plant located next to a reservation. This coal-burning power plant had been dumping toxic waste on the reservation's residents for decades and was the object of longstanding protests (see Solar on Tribal Lands section below).

This goes back to the bathtub metaphor. By agreeing to purchase solar power, Las Vegas generated demand to make the total energy mix a little bit cleaner. The electricity for the City of Las Vegas actually comes from sources produced throughout the state because, as shown in the bathtub metaphor, all the electrons are the same once they are in the grid. But, because they agreed to purchase solar for their electrical use, NV Energy had the authorization to add more solar power to the bathtub. As a city, Las Vegas used its buying power as an economic driver to reduce pollution. As an investor-owned utility, NV Energy needs the purchase power agreements to build new means of production. With the Boulder City solar plant online, NV Energy did not need the energy from this coal-burning power plant any longer.

To know how much they had to buy, the City of Las Vegas had to do extensive research to get an accurate assessment of their retail load. This research extended back several years to see the growth in electrical use, particularly in consideration of the dynamic population in Las Vegas. Once the city established this number, they worked with NV Energy to determine how much solar they had to purchase to cover their use.

In speaking with the city planners, they emphasized that solar was just one part of the sustainability measures. Other elements, like installing LED lights, were also critical to the success of their plan, as LED lights reduce the retail load, so there is less power being used. Renewable energy does not solve underlying issues of energy waste, and simply switching to newer LED technology creates infrastructure that is far more efficient to operate and requires less power. Working in multiple ways, to conserve resources and use them as efficiently as possible, is as important to sustainability as installing solar panels.

Combining renewable energy credits, city officials said they reduced their overall utility costs by a third, from about 15 million dollars annually, in 2008, to less than 10 million dollars per year in 2016.

Higgins Generating Station

To understand the technical aspects of how a power generation facility works, NV Energy invited us to film at one of their power plants. Higgins Generating Station is located 45 miles south of Las Vegas, on the California state line. It is a 530-megawatt natural gas-burning power plant, surrounded by the Silver State North and South Solar farms, which combined are over 300 megawatts on 2,900 acres.[21] Within this expanse of the sparse Mojave desert, we can see how the technology of this type of system works.

In interviews with executives and engineers from NV Energy, they explained how the various pieces of the grid work together. With the population of the state growing, there is a need for more electricity around the clock. Solar produces electricity when the sun is shining, in the middle of the day. However, the grid operates 24 hours a day, 365 days a year. As mentioned in chapter 2, one of the essential physical aspects of electricity is that it is instant. As soon as it is produced, it is delivered to your home. Unlike a physical resource like water, which can be stored in a reservoir, the intangible nature of electricity makes it a challenge to store.

Once the sun hits the solar panel at Higgins Generating Station, or Nellis Air Force Base, or panels outside Las Vegas City Hall, or on a tribal reservation, the electrons flow through wires, along the grid, and to a building to run an air conditioner, charge a cell phone, or turn on the lights in an elementary school. The utility needs to ensure that they can match that same amount of electricity when the sun is not shining and other types of renewable energy are not producing: at night, on cloudy days, or other times when the sun is not shining.

The engineer from NV Energy gave our film crew a tour, showing us the various systems and taking us through the power plant. He explained the basics, including the duck curve. This is where the science of energy distribution kicks in. Natural resources still limit even a utility that very much wants to use solar or wind. The major advantage that fossil fuel power plants have is that you can control their output, and they run all the time. Clearly, they also have major disadvantages, from toxic waste to energy dependence.

The engineer also shared that Nevada has a "renewables first" policy, and whenever power from solar or other renewable energy is available, it is used first. One of the major challenges that utilities currently face is how to create storage systems that balance the electrical production from solar and other renewable energy, so it can be delivered all day and into the night. (Many other chapters in this book address energy storage and battery use by utilities and homes.)

Fossil fuel-burning power plants are relatively simple technology. They burn the fuel– coal, oil, natural gas—which creates steam and which spins a turbine. The spinning of the turbine creates electricity. Utilities are required by law to produce more electricity than the public needs at any given time. This is because, in the event that there is a surge in demand, utilities need to have the power to immediately meet that demand. If you have ever experienced a "brown-out," that is when the utility is not able to meet the amount of electricity being used by its customers, for example, on extremely hot days when everyone's air conditioners are on.

Throughout the U.S., systems like NV Energy, which use many different types of electrical production to feed into 1 grid, have introduced system operators. While this is a simplification of a more complex process, essentially, the system operator watches the demand through monitors, watches the supply, and coordinates the input of various sources of electricity into the system. Without this overall coordination, too much energy could flood the system, causing the surges and possible explosions detailed in chapter 2.

Energy Storage

At the time of this interview (2017), storage for electricity had only barely started to come to market. At the time of finishing this book, many utilities have begun using batteries, testing them to see how they can function as part of this grid system. Just like any other battery, a utility-scale battery can be charged up for when it is needed. On a solar power system of any size, whether a utility-scale or small emergency-sized system, the battery charges during the day and kicks in when the duck curve hits the high point of demand in the late evening when people come home from work.

This is why battery technology is so important to clean energy systems—they create the same kind of reservoir we use for water, allowing the system to be more stable and for the solar power to be extended from noon to late at night. Or for wind power to be stored from days when winds blow more consistently until whenever that power is needed.

It is clear how batteries are a crucial piece of the puzzle, and hopefully, by the time you are reading this book, energy storage will be as ubiquitous as solar is becoming today. But, until then, utilities must anticipate the duck curve of demand and create electricity.

LARGER STATEWIDE PICTURE

Throughout this chapter, we have seen the solar consumption of Las Vegas and the different city initiatives the city took to meet its clean energy goals.

However, the conversation on solar in Nevada would be incomplete without also discussing the solar is being used by other entities, as well as a threat very specific to Nevada.

Residential net metering

Residential solar and a statewide, citizen-led campaign for access to affordable solar was also not able to be included in the scope of this research but were very influential for solar. At the same time that I was interviewing with NV Energy, a statewide campaign about net metering had resulted in a legislative battle. The capacity of net metering to engage the population in integrating solar and powering our cities with clean energy is enormous, simply because it gives those families a share of the economic benefits of solar. In a state with the amount of sunshine that Nevada has, you can imagine that a family can take great advantage of the capacity of a solar power system to reduce their own electric bill.

The citizen-led campaign centered on the rate the utility was paying a household for selling electricity back into the grid. NV Energy had proposed to lower the amount that a consumer would be paid for this, which significantly damaged the fast-growing Nevada solar market. The citizen campaign fought to keep the price at a much higher level, allowing a family to invest in a solar power system, pay it off in just a few years, and have the same kind of an extremely low electricity bill as the Staten Island family discussed in chapter 2, who had a $17 per month electric bill due to the net metering policies. Nevadans also had very good net metering policies, making it affordable to purchase solar power systems and see them pay themselves off in just a few years.

This is another area where my research did not allow for more inquiry; however, ultimately, the citizen campaign won. The mainstream press and academic studies have fully detailed both sides of this controversy, the ensuing campaign, utility hearings, legislative decision, and the victory, which reestablished consumer-friendly rates for net metering.[22] The bill that the Nevada government passed was seen as a huge victory for solar advocates. It established that Nevadans have the right to install rooftop solar, connect the panels to the grid, and store power in batteries (and thus address the intermittency issue, so they can save more money). Researchers continue to look at this citizen campaign, as well as net metering in general, and how this very powerful tool impacts clean energy possibilities.[23,24]

Consumer access and net metering are absolutely crucial aspects of how any city or stature will power itself. Ensuring that all families can benefit from the cost-saving and resiliency measures of solar is the only way to realize the full potential of renewable energy.

Geothermal Utility Power

NV Energy has a sizable amount of geothermal—the second highest in the country for utility geothermal. It is one of a small handful of states to be able to use this type of renewable energy at a utility-scale, although geothermal for residences and businesses is used throughout the country. Geothermal systems are pipes that go down several feet and use the natural heat below ground to heat water or another liquid and create energy.

At the time of filming, NV Energy had 19 geothermal projects on the grid, mostly located in the northern part of the state. While I did not film at any of these, it is worth mentioning as another clean, renewable energy source.

Solar on Tribal Lands

Three months after the City of Las Vegas made the announcement about reaching 100 percent renewable energy, NV Energy shut down the second-to-last coal-burning power plant in the state, the Reid Gardner Generating Station. Large contracts for solar development, like a city-sized one, allowed NV Energy to build solar plants and meet not only statewide electricity demand but the specific demand from the City of Las Vegas through the solar contracts.

Las Vegas made its announcement about achieving 100 percent renewable energy in December 2016. In February 2017, NV Energy flipped the switch on a new, 50megawatt solar power plant, Boulder Solar II.[25] In March 2017, they flipped the switch off on the aging Reid Gardner Generating Station, which was located next to the lands of the Moapa Band of Paiutes. The tribe had long held the 50-plus-year-old power plant responsible for respiratory illnesses, cancers, and asthma among tribal members.[26]

Shutting the Reid Gardner plant was no small act, and it can be attributed to years of activism demanding it be shut down, including a protest on Earth Day 2012 in which tribal members walked from the plant to the federal building in Las Vegas for 3 days of protests about the plant's toxic waste.

The tribal protest against the power plant deserves dedicated study, and unfortunately, falls outside of the scope of my research. However, the conclusion is that the Moapa Band of Paiutes signed a lease to allow a solar farm to be built on their lands, which opened the day after the Reid Gardner plant shut down in 2017. The Moapa Southern Paiute Solar Project provides jobs for over 100 tribal members, as well as a revenue stream for the tribe through the lease of their land, and created a means for the tribe to also be part of a solar economy.[27]

As solar technology comes into grids and new clean energy power production sources are built, aging and polluting sources like old coal-burning power plants can be shut down. The many solar developments throughout Nevada create an ecosystem of clean energy technology, business models, and stakeholders, allowing this transition.

Nellis Air Force Base

Nellis Air Force Base is less than 10 miles away from Las Vegas. In addition to city governments, residents, casinos, private businesses, and tribal authorities, the solar ecosystem also includes the military. When asked if my film crew could film at one of their solar installations, NV Energy was able to get access for us to go to Nellis Air Force Base. It is considered a solar success story for many reasons. First of all, it creates resiliency and sustainability for the air force base. The original construction on Nellis AFB was designed to supply about 25 percent of the base's electricity. It came online in 2007. Through purchase power agreements, the U.S. Air Force saved over 1 million dollars in energy costs and reduced greenhouse gas contributions by 24,000 tons. And, they repurposed land that is not suitable for residences—a former landfill—for the solar array.[28]

Called the Nellis Solar Array II Generating Station, this is the only solar power station in all of the locations where I filmed, where there was a need to clean the solar panels. Generally, most people have told me maintenance of solar farms is minimal and cleaning them is rarely necessary. Because desert winds blow fine dust that settles on the panels, reducing their effectiveness, here, they needed cleaning. With over 70,000 panels, the Nellis array has a specially designed set of robot cleaners, which use one-quarter of the water of manual cleaning.[29]

NV Energy executives said Nellis has since expanded from the original amount of solar panels and is now a 114-megawatt site. The first one, in 2007, was the largest of its kind in the world, at that time, on a military base. With the expansion, at the time of filming in 2017, Nellis was the only military installation in the world that generated more electricity than it used on an annual basis.

Casinos and Solar

As mentioned above, at the time of filming, the Mandalay Bay casino had installed an 8.3 megawatts rooftop solar system, which was regarded as the largest in the United States at that time.[30] The Mandalay Bay Casino also won a potentially influential legal victory, which is to pay an "impact fee," more commonly referred to as an "exit fee," that would allow them to separate from the grid and no longer buy electricity from the utility.[31] This type of fee could potentially become a larger issue down the road, particularly for very large solar installations and private businesses. A grid functioning as described in this chapter depends on everyone using it to have the economy of scale to make it affordable. However, some businesses are looking for other ways to create their own electricity. Casinos, with their large rooftops and the resources to build massive solar-and-battery systems, seem to have found one.

The exit fee paid by MGM, which owns the Mandalay Bay casino, was $87 million. That seems like a lot of money, and perhaps it is. But not for a Vegas casino. It is roughly the equivalent of what they would have paid the utility for 1 year of electricity.

This same concept of leaving the grid came up in research conversations in Puerto Rico. Similarly, businesses looked to build microgrids or large solar-and-battery systems, and while they do not all have the resources of a Vegas casino, some of the hotels, resorts, and corporations have the means to invest in this. It is an extremely new development and one that researchers on energy systems should follow.

MGM is not the only casino looking to leave the utility. In 2019, a news report cited over 10 other casinos that have filed paperwork to leave the utility. While I was not able to expand my research to look at casinos and exit fees, it bears mentioning as this could have a tremendous influence on cities going 100 percent renewable energy.

Nuclear Dumping

There is one final energy-related story that is pertinent to Nevada and bears mention in any discussion of clean energy: nuclear dumping. While net metering and solar have divided Nevadans, they are firmly united in their opposition to becoming the nation's nuclear waste dump. If the opposition to nuclear dumping in Nevada had a de-facto leader, it would be Mayor Carolyn Goodman.

How does this tie into clean, renewable energy? Yucca Mountain lies about 90 miles from Las Vegas and has been proposed as a site for a nuclear waste repository, which is a fancy term for a waste dump. As the focus of this book is on community responses to energy-related issues, and the readers of this book are future or current professionals in energy, then it is important to address issues that do not always get the same attention as solar and wind development.

Nuclear energy is not discussed throughout the rest of this book because in reaching out to people working on projects to transform energy in their communities, no one wanted to bring a nuclear power plant to their community. In looking at broader energy issues in Nevada, a looming threat emerged in a national plan to make Nevada a dump site for the nation's nuclear waste: Yucca Mountain.

Nuclear power is considered clean energy because it does not produce carbon emissions. However, it does produce radioactive waste. Finding a way to manage this radioactive waste has been a national priority since the early 1980s, supported by billions of dollars of federal funding, which was authorized through the Nuclear Waste Policy Act, more colorfully referred to by many Nevadans as the "Screw Nevada" Law. For over 30 years, the federally funded Yucca Mountain site has been developed as a potential place to truck all of the nation's nuclear waste and dump it. For 30 years, Nevada leaders from all political corners have opposed this plan. Funding for Yucca Mountain was cut a decade ago, during the Obama Administration, but they never stopped it legally.[32] Under the Trump administration, funding for Yucca Mountain was restarted.[33]

Mayor Goodman's and NV Energy's initiatives on solar and clean energy are seen with even more clarity in light of the forceful, longstanding, vocal opposition to nuclear waste from across Nevada, which produces zero nuclear waste, yet is at risk of becoming a radioactive waste dump.

The Yucca Mountain proposal began under the Reagan administration. It has received billions in federal dollars to study and design how to build an underground facility in which we could bury all the radioactive nuclear waste from across the nation. Currently, this waste is buried onsite at the power plants where it was created. In order for this plan to work, the radioactive waste would have to be loaded onto trucks and trains and transported for thousands of miles across the entire country. For this reason, hundreds of mayors across the country have joined Nevada's elected officials in opposing Yucca Mountain.

Even with all their clean energy sources being added to the grid, Nevadans fear that Yucca Mountain might resurrect itself again. In 2020, an editorial in the Las Vegas Sun called Yucca Mountain "unfair, unsafe, and undead,"[34] and detailed how Nevada's elected officials must continue to oppose this ill-fated proposal, which manages to remain alive at the Nuclear Regulatory Committee.

Nevada has no nuclear power plants. It ranks high every year in multiple metrics for progress on solar and renewable energy. It is reducing pollution by closing coal-burning fossil fuel power plants. Mayor Goodman has made a tradition of speaking out against Yucca Mountain, as did her husband before her, with widespread support from others in the U.S. Conference of Mayors.

Any reader with a serious interest in energy development should look more into Yucca Mountain and ask these questions: Why would a state that has no nuclear power plants win the extremely dubious position as a proposed site for all of the nation's nuclear waste? What are the community health impacts of creating a large repository for nuclear waste? What are the dangers of shipping nuclear waste on trains and trucks around the country? If Nevada effectively stops this repository, where might it go next?

An excellent resource for more background is the Nevada Commission on Nuclear Projects' Report and Recommendations to the state legislature, which is an exhaustive overview of the history of Yucca Mountain.[35]

CONCLUSION

This chapter began with a question about what insights can Las Vegas and Nevada yield for other places. It closes with a conclusion about how a clear sense of direction from the Mayor of Las Vegas, including a commitment to energy issues that were begun by the preceding mayor. Continuity like this was one reason that the City of Las Vegas was able to hit the 100 percent renewable energy mark for its city services. They began in 2006 and made their announcement about being 100 percent renewable at the end of 2016—a decade-long process of transformation. As other cities move toward this milestone, it would be valuable to know if all cities will take 10 years and whether unified political goals speed up or hinder their transition. This chapter also closes with a look at the broader energy issues, and how the city's efforts intersect with many other equally-important issues ranging from tribal lands and energy independence to nuclear waste storage.

The access that the City of Las Vegas officials and NV Energy provided and the information they shared are something that all government and utility officials should consider as important in a public communications program about city-based renewable energy strategies. It is well-documented that people across the U.S. want renewable energy. However, very few people know what their city is doing or how they can be a part of those efforts. A city and state that is serious about achieving a goal of 100 percent percent renewable energy needs to also ask how they can best educate the public and engage city and state residents in expanding renewable energy use and efficiency.

Another lesson of this chapter is an unfortunate one—even if the state and city are making tremendous strides, they still face risks from outside sources of pollution. The essence of energy is its interconnection. As long as our nation produces toxic and radioactive wastes, we will need to find places to store them.

Oh, and that pressing heat? It is going to get worse. In 2019, Climate Central used decades of weather reports to crunch the data. Guess which city is the fastest-warming city in America? If you guessed Las Vegas, you would win. Measuring average temperatures from 1970– 2018, the national average temperature climbed 2.5 degrees. In Las Vegas, it hiked up 5.7 degrees.[36] So, not only are all those air conditioners going to stay on in the near future, but they will become critical infrastructure that is necessary for human survival.

NOTES

1. C40 Cities Report. "CDP Cities 2013: Summary Report on 110 Global Cities." *C40 Cities Climate Leadership Group, C40.org*. CDP. 2013 https://www.c40.org/researches/c40-cdp-2013-summary-report

2. National Oceanic and Atmospheric Administration, Ranking of Cities Based on Percentage of Annual Possible Sunshine. https://www1.ncdc.noaa.gov/pub/data/ccd-data/pctposrank.txt

3. City of Las Vegas, *Sustainability Reports*, Accessed on January 1, 2021 https://www.lasvegasnevada.gov/Government/Initiatives/Sustainability/Sustainability-Resources?tab=0

4. Martinez, Hayley, DeFrancia, Kelsie, and Schroder, Alix. "Moving Towards 100% Renewable Energy: Drivers Behind City Policies and Pledges." Synopsis published on *State of The Planet, Earth Institute*, April 24, 2018. Full report presented at the 76th Annual Midwest Political Science Association Conference, 2018. https://blogs.ei.columbia.edu/2018/04/24/moving-towards-100-renewable-energy-drivers-behind-city-policies-pledges/

5. https://wwwusmayors.org/programs/mayors-climate-protection-center/.

6. Fialka, John. "As Hawaii Aims for 100% Renewable Energy, Other States Watching Closely." *Scientific American*, April 27, 2018. https://www.scientificamerican.com/article/as-hawaii-aims-for-100-renewable-energy-other-states-watching-closely/#:~:text=When%20it%20comes%20to%20generating,percent%20renewable%20energy%20by%202045.

7. Shahan, Zachary. 2020. "Top US States for Percentage of Electricity from Solar." *CleanTechnica*. October 5, 2020. https://cleantechnica.com/2020/10/05/top-us-states-for-percentage-of-electricity-from-solar-cleantechnica-report/.

8. Shahan, "Top US States for Percentage of Solar."

9. Morehouse, Catherine. "Nevada Passes Bill for 50% Renewables by 2030, 100% Carbon Free by 2050." *Utility Dive*, April 22, 2109.

10. SEIA/Wood Mackenzie Power & Renewables U.S. Solar Market Insight 2020 Q4, Solar Industry Research Data. Accessed January 2, 2021. https://www.seia.org/solar-industry-research-data

11. Luskin Center for Innovation. *Progress Toward 100% Clean Energy in Cities and States Across the U.S.* UCLA Luskin Center for Innovation (LCI). November 6, 2019. https://innovation.luskin.ucla.edu/wp-content/uploads/2019/11/100-Clean-Energy-Progress-Report-UCLA-2.pdf

12. Pothecary, Sam. "The Largest Solar Rooftop System in the US Installed on the Mandalay Bay Convention Center." *PV Magazine*, July 7, 2016.

13. Chow, Lorraine. 2016. "Kansas Town Decimated by Tornado Now Runs on 100% Renewable Energy, Should Be Model for Frack-Happy State." *EcoWatch.com*. October 18, 2016. https://www.ecowatch.com/greensburg-kansas-renewable-energy-2051037395.html.

14. U.S. Department of Energy. "Confronting the Duck Curve: How to Address Over-Generation of Solar Energy." *Office of Energy Efficiency & Renewable Energy*, October 12, 2017. https://www.energy.gov/eere/articles/confronting-duck-curve-how-address-over-generation-solar-energy

15. Toplikar, Dave. "New Las Vegas City Hall Dedication a Highlight of Downtown Development." *Las Vegas Sun*, Monday, March 5, 2012. https://lasvegassun.com/news/2012/mar/05/new-city-hall-dedicated-light-show/

16. U.S. Census data - https://www.census.gov/quickfacts/lasvegascitynevada and Paradise census data - https://www.census.gov/quickfacts/paradisecdpnevada and visitor stats (including pre-COVID) - https://www.lvcva.com/research/visitor-statistics/

17. U.S. Conference of Mayors, "Mayors Climate Protection Agreement." Resolution passed in 2005. Accessed January 2, 2021. https://wwwusmayors.org/programs/mayors-climate-protection-center/.

18. Hernandez, Dan. 2016. "Behind the Bright Lights of Vegas: How the 24-Hour Party City Is Greening up Its Act." *The Guardian*. May 3, 2016. https://www.theguardian.com/sustainable-business/2016/may/03/las-vegas-strip-sustainability-mgm-caesars-water-resource-drought.

19. Green Chips Clark County, NV 2016 report - http://www.clarkcountynv.gov/comprehensive-planning/eco-county/announcements/Pages/Green-Chips-Southern-Nevada-State-of-Sustainability-Report-2016.aspx; Las Vegas sustainability initiatives -https://www.lasvegasnevada.gov/Government/Initiatives/Sustainability

20. Vogel, Ed. "NV ENERGY Launches Discount LED Lighting Program for Consumers." *Las Vegas Review-Journal*, February 19, 2014. https://www.reviewjournal.com/news/nv-energy-launches-discount-led-lighting-program-for-consumers/

21. U.S. Department of Energy. "DOI Announces Two Solar Projects Approved in California, Nevada." *Department of Energy* website, February 26, 2014. https://www.energy.gov/eere/solar/articles/doi-announces-two-solar-projects-approved-california-nevada

22. Shogran, Elizabeth. "In Solar Scuffle, Big Utilities Meet Their Match." *High Country News*, August 21, 2017. https://www.hcn.org/issues/49.14/solar-energy-solar-eclipse-big-utilities-meet-their-match-in-solar-scuffle

23. Burgy, Robert S. "A Dramatistic Analysis of Nevada's Controversy over Solar Net Metering Incentive Policies." UNLV Theses, Dissertations, Professional Papers, and Capstones, 2018: 3225.doi: 10.34917/13568399

24. Davies, Lincoln L., and Sanya Carley. "Emerging Shadows in National Solar Policy? Nevada's Net Metering Transition in Context." *The Electricity Journal* 30, no. 1 (2017): 33–42. https://www.sciencedirect.com/science/article/abs/pii/S1040619016301762

25. KTNV Channel 13 Las Vegas. "New Solar Plant Near Boulder City." *KTNV Channel* 13 Las Vegas Youtube channel, February 14, 2017. https://www.youtube.com/watch?v=lZ5N6Ne0TSw&ab_channel=KTNVChannel13LasVegas

26. Ritter, Ken. "NV Energy Pulls Plug on Coal-fired Power Plant Near Las Vegas." *Las Vegas Sun*, March 16, 2017.

27. Funes, Yessenia. "How One Small Tribe Beat Coal and Built a Solar Plant.". *Colorlines.Com*. 2017. https://www.colorlines.com/articles/how-one-small-tribe-beat-coal-and-built-solar-plant.

28. U.S. Environmental Protection Agency. "Re-Powering America's Land: Siting Renewable Energy on Potentially Contaminated Land and Mine Sites, Nellis Air Force Base, Nevada Success Story." February 2009. https://www.epa.gov/sites/production/files/2015-04/documents/success_nellis_nv.pdf

29. NV Energy. "Nellis Solar Array II Generating Station flyer." Website accessed January 2, 2021. https://www.nvenergy.com/publish/content/dam/nvenergy/brochures_arch/about-nvenergy/our-company/power-supply/Nellis-Fact-Sheet.pdf

30. Brean, Henry. "MGM Resorts Will Use Solar Arrays to Power Las Vegas Casinos." *Las Vegas Review-Journal,* April 18, 2018

31. Spector, Julian. "How MGM Prepared Itself to Leave Nevada's Biggest Utility: The Casino Conglomerate Expects to Double Its Use of Renewable Energy and Earn Payback Within 7 Years." *Greentech Media*, September 16, 2016.

32. Zhang, Sarah. "The White House Revives a Controversial Plan for Nuclear Waste: Yucca Mountain is Back, and Nevadans are Not Happy." *The Atlantic,* March 21, 2017. https://www.theatlantic.com/science/archive/2017/03/yucca-mountain-trump/519972/

33. Maher, Kris. "Trump Administration Revives Nevada Plan as Nuclear Waste Piles Up." *Wall Street Journal,* May 9, 2017. https://www.wsj.com/articles/a-glimmer-of-hope-in-a-growing-hoard-of-nuclear-waste-1494345622

34. Halstead, Bob. "Nevada Is Winning the War, but Yucca Mountain Is Not Dead." *Las Vegas Sun*, August 23, 2020. https://lasvegassun.com/news/2020/aug/23/nevada-is-winning-the-war-but-yucca-mountain-is-no/

35. Nevada Commission on Nuclear Projects. "Report and Recommendations, November 2019." Www.State.NV.us/Nucwaste/. State of Nevada. November 2019. http://www.state.nv.us/nucwaste/pdf/2019%20Commission%20Report%20and%20Recommendations.pdf

36. Climate Central. "American Warming: The Fastest-Warming Cities and States in the U.S." *ClimateCentral.org.* April 17, 2019. https://www.climatecentral.org/news/report-american-warming-us-heats-up-earth-day

Chapter 11

The End? or Another Beginning?

As this book goes to press, the coronavirus pandemic that gripped the United States has finally started to diminish as vaccines become available, however, it remains to be seen whether new variants will create more surges. Covid-19 also continues to spread in countries around the world, with less access to vaccines and treatment. As the U.S. slowly starts to resume pre-pandemic activities, communities are also coping with the havoc it inflicted on public health and the economy. The pandemic and international lockdowns revealed a related energy vulnerability with our system: our near-complete dependence on electricity.

The United States already had an enormous electrical demand: the second-highest in the world. During the lockdown, as remote school, remote work, remote doctor appointments, remote meetings, and nearly every other aspect of daily life went online, what happened to people who lost power? If Hurricanes Sandy, Maria, and many others revealed the extreme flaws in our energy system before the pandemic, then the lockdown revealed our equally extreme need for constant, reliable electricity. We can now see the sizable stakes if we do not fix these flaws, as the pandemic showed us that reliable power is a matter of public health. We can also see several ways individuals, communities, and whole cities are starting to enact solutions.

Along the road of filming and editing the documentary and writing this book, I saw how complicated energy transformation will be when looked at from a macro scale. I also saw how simple it is when approached from a household level. One of the significant challenges to transition is that most people simply cannot see our energy system. I showed a rough scene of the film to a colleague, with a shot of 1 of Astoria's power plants dominating the skyline. My colleague commented that he simply did not "see" the power plant, nor did he question its location amid residential homes. That it was merely what New York City's skyline looks like. This plant is just one of many which dot our horizons, so ordinary that few people question why they are close to residential homes. Fossil fuel-burning power plants have been part of the structure, outline, and image of our cities and towns for over a century. This familiarity of power systems makes them invisible—and also makes people think they are immutable. It is impossible to change something that you barely see. Starting to see the pieces of our energy system is one of the first steps in replacing those pieces, removing the unnecessary ones, and transforming the whole thing to a better design.

Clean energy projects like those profiled throughout this book, started by individuals, governments, and community organizations, show that our power system is quite changeable. Excellent historical accounts of our power system describe evolution like moving from whale oil to steam power and fossil fuels, developing the grid, and creating an electrification system that reaches 100 percent of homes in the United States.[1,2] Those changes are starting to include solar, offshore wind, geothermal power, and other types of clean and renewable energy, as well as how the grid can evolve to something better.

Transformation of our energy system allows us to take advantage of the benefits of distributed generation, solar power, offshore wind turbines, batteries, geothermal, and many types of clean, renewable energy. The challenges are not just political or economic, and they are most definitely not technical. The challenges also include developing public trust and building knowledge and understanding about how these systems work. While these are massive transitions, they are not more significant than other major technology transitions that also involved cultural change. For example, going from horses to cars, landlines to cell phones, and watching broadcast television to streaming social media content, all involved parallel changes in trust, knowledge, culture, and consumer behavior.

As reflected in the quotes throughout this book, success in energy, sustainability, solar, policy, and environmental communications also requires an ability to maintain optimism, finding a sense of connection to one's community, and approach the challenges of energy transition with creativity. The ACUA was on the cutting edge of technology in the late 2000s when they installed their wind turbines. Now they have a substantial clean energy system with multiple generation sources and battery storage that creates 100 percent of the power they consume. When Green Mountain Power first offered Mary Powell a job, she turned it down—but ultimately joined the company and transformed it. In Puerto Rico, an island-wide movement uses solar power in multiple ways, training young people to become solar leaders, and applying the massive influx of solar technology that arrived after Hurricane Maria to improve community safety and provide economic benefit. In Highland Park, a community mobilized, educated themselves, and became part of a statewide initiative that advanced renewable energy policies. Las Vegas, and all of Nevada, have been quietly leading the way for decades and are already forging new terrain for clean energy. And, New York City has already started on its course to transformation, from cleaning up dirty buildings to passing a law joining restorative justice with clean energy.

Even though the research that led to this book and the related film came to an end, all of the projects and people portrayed in these chapters continue. They show us what is around the corner. A Green New Deal for our nation's largest city. Cities using 100 percent renewable energy for their buildings and services. States mandating 100 percent renewable energy goals. Community education engaging people in determining their energy future. In some places, utilities are moving from "twigs and twine" to batteries and distributed generation. In other places, corporations are walking away from the utilities. Power systems can now attach batteries, creating adaptable on-and off-grid power systems. Training programs are putting skills and knowledge in the hands of those most affected by climate change and most able to use 100 percent solar, creating excellent models for expanding training. Offshore wind is just on the horizon, as is battery storage for whole utilities.

All of these are not the end. They are the next beginnings.

NOTES

1. Bakke, Gretchen. *The Grid: The Fraying Wires Between Americans and Our Energy Future*, 1st edition. *Bloomsbury USA*, July 26, 2016.

2. Rhodes, Richard. *Energy: A Human History*, 1st ed. New York: Simon & Schuster, 2018.

Additional Resources

Hawken, Paul, and Wilkinson, Katherine. *Drawdown: The Most Comprehensive Plan Ever Proposed to Reverse Global Warming.* Edited by Paul Hawken. *Amazon.* New York: Penguin, 2017. https://drawdown.org/the-book.

Johnson, Ayana Elizabeth, and Wilkinson, Katharine K, eds. *All We Can Save: Truth, Courage, and Solutions for the Climate Crisis. Amazon.* New York: Random House, 2020. https://www.allwecansave.earth/.

Kolbert, Elizabeth. *The Sixth Extinction.* London, UK: Bloomsbury Publishing PLC. 2015.

Mulvaney, Dustin. *Solar Power: Innovation, Sustainability, and Environmental Justice.* Oakland, CA: University of California Press, 2019.

Onís, Catalina M de. *Energy Islands: Metaphors of Power, Extractivism, and Justice in Puerto Rico.* Oakland, CA: University Of California Press, 2021.

Watt-Cloutier, Sheila, and McKibben, Bill. *The Right to Be Cold: One Woman's Fight to Protect the Arctic and Save the Planet from Climate Change. Project MUSE.* Minneapolis, MN: University of Minnesota Press, 2018.

PODCASTS

"Decolonizing Power." Bi-weekly Podcast, Mihskakwan James Harper and Freddie Huppé Campbell, hosts. Indigenous Clean Energy Network: 2021. https://anchor.fm/decolonizing-power

"The Energy Gang." Weekly Podcast. Lacey, Stephen, Katherine Hamilton, and Jigar Shah, hosts. Greentech Media: 2013–2021. https://www.greentechmedia.com/podcast/the-energy-gang.

"Think 100%: The Coolest Show." Weekly Podcast. Lacey, Stephen, Katherine Hamilton, and Jigar Shah, hosts. Greentech Media: 2013–2021. https://think100climate.com/podcasts/coolest-show-on-climate-change.

"Watt It Takes." Weekly Podcast, Emily Kirsch, host. Powerhouse: 2021 https://www.powerhouse.fund/wattittakes.

Bibliography

Alemazkoor, N., Rachunok, B., Chavas, D.R. et al. "Hurricane-induced Power Outage Risk Under Climate Change Is Primarily Driven by the Uncertainty in Projections of Future Hurricane Frequency." *Science Reports* 10 (2020), 15270. doi: 10.1038/s41598-020-72207-z.

Alfonso, Omar. "Arroyo Barril: Coal Ash and Death Remain 15 Years Later." *Periodismo Investigativo,* December 20, 2018.Azzopardi, Tom. "AES Corp. Fined $5 Million in Chile Over Coal Plant Data." *Bloomberg Law*, June 27, 2019. https://news.bloomberglaw.com/environment-and-energy/aes-corp-fined-5-million-in-chile-over-coal-misstatement.

Alfonso, Omar. "Damage by Coal Ash to the Southern Aquifer Cannot be Undone." *Center for Investigative Journalism.* March 25, 2019.

Alfonso, Omar. "EPA Adopts New Rules Tailored for AES Coal Plant Puerto Rico." *La Perla del Sur y Centro de Periodismo Investigativo.* August 3, 2018.

American Eagle Foundation. "Promote Avian-Friendly Power Lines: Millions of Birds Fatally Collide with and Are Electrocuted by Power Lines Annually. There Are Ways to Prevent This!" *Eagles.org.* Accessed May 15, 2021. https://www.eagles.org/take-action/avian-friendly-power-lines/.

ANDA Asociación Nacional de Derecho Ambiental and Enlace Latino de Acción Climática. "Event Listing: Transformación Energética Desde Las Comunidades (Community-Based Energy Transformation)." December 7, 2018. https://www.facebook.com/events/286994762144014/.

Aronoff, Kate. "Armed Federal Agents Enter Warehouse in Puerto Rico to Seize Hoarded Electric Equipment." *The Intercept.* 2018. https://theintercept.com/2018/01/10/puerto-rico-electricity-prepa-hurricane-maria/.

Art-Ops. "Community heART Highland Park." *Art-Ops.org*, accessed December 14, 2021. https://art-ops.org/community-arts-highland-park.

Ashoka. "Social Entrepreneurship." *Ashoka.org*, accessed January 2, 2021. https://www.ashoka.org/en-us/focus/social-entrepreneurship.

Atlanta Electrician. "White Paper: What is Electrical Arcing and Why is it Dangerous?" Atlanta Electrician, *Atlanta.com.* December 28, 2017. https://www.electricianatlanta.net/what-is-electrical-arcing-and-why-is-it-dangerous/.

Atlantic County Utility Authority. "Jersey-Atlantic Wind Farm." *ACUA.com,* accessed December 29, 2020. http://www.acua.com/green-initiatives/renewable-energy/windfarm/.

Auffhammer, M. "Do Black Households in the U.S. Pay More for Their Energy?" *Energy Post.* Last modified July 23, 2020. https://energypost.eu/do-black-households-in-the-u-s-pay-more-for-their-energy/.

Azzopardi, Tom. "AES Corp. Fined $5 Million in Chile Over Coal Plant Data." *Bloomberg Law,* June 27, 2019. https://news.bloomberglaw.com/environment-and-energy/aes-corp-fined-5-million-in-chile-over-coal-misstatement.

B-Labs. "About B-Corps." *B-corporation website,* accessed January 2, 2021. https://bcorporation.net/about-b-corps.

Bagley, Katherine. "After The Storm, Puerto Rico Misses Chance to Rebuild With Renewables." *Yale E360,* May 31, 2018. https://e360.yale.edu/features/after-the-storm-puerto-rico-misses-a-chance-to-rebuild-with-renewables-hurricane-maria.

Bakke, Gretchen. *The Grid: The Fraying Wires Between Americans and Our Energy Future,* 1st edition. Bloomsbury USA, July 26, 2016.

Bednar, Dominic J., and Tony G. Reames. "Recognition of and Response to Energy Poverty in the United States." *Nature Energy* (2020): 1–8. https://www.nature.com/articles/s41560-020-0582-0?draft=collection.

Before the Senate Committee on Energy and Natural Resources." *U.S. House of Representatives, 105th Congress, 1st Session, Committee On Natural Resources,* Tuesday, November 14, 2017. Full Hearing Text: https://www.govinfo.gov/content/pkg/CHRG-115hhrg27587/html/CHRG-115hhrg27587.htm Gov. Mapp's Written Statement: https://www.energy.senate.gov/services/files/A2538A49-2953-4BA1-8C94-0807E62050A5.

Binelli, Mark. *Detroit City Is the Place to Be: The Afterlife of an American Metropolis.* New York: Picador, 2013.

Blue Planet Energy. "Press Release: Blue Planet Energy Joins Puerto Rico Electrification Efforts With Battery Deployments and Trainings." *Pv-magazine-usa.com,* March 2018. https://pv-magazine-usa.com/press-releases/blue-planet-energy-joins-puerto-rico-electrification-efforts-with-battery-deployments-and-trainings/.

Blue Planet Energy. "Water Mission Delivers Clean Water to over 1,200 People with Blue Ion in Puerto Rico - Blue Planet Energy." *BluePlanetEnergy.com,* October 30, 2018. https://www.blueplanetenergy.com/water-mission-delivers-clean-water-with-clean-energy-in-puerto-rico/.

Blumsack, Seth. "Introduction to the Electricity Industry: Electricity Demand and Supply in the United States, EME 801: Energy Markets, Policy, and Regulation." Penn State, *Department of Energy and Mineral Engineering,* website accessed December 12, 2020. https://www.e-education.psu.edu/eme801/node/490.

Bohr, Jeremiah, and Anna C. McCreery. "Do Energy Burdens Contribute to Economic Poverty in the United States? a Panel Analysis." *Social Forces* 99, no. 1 (2020): 155–177. doi: 10.1093/sf/soz131.

Bolger, Kyle. "How to Power a City." Interview by Melanie La Rosa. March, 2018.

Bolstad, Jennifer. "How to Power a City." Interview by Melanie La Rosa. January 2018.

Bratspies, Rebecca M. "Public Housing, Private Owners: Sustainable Development Lessons from the Fight to Shut the Poletti Power Plant." *Papers.ssrn.com.* Rochester, NY. May 6, 2020. https://ssrn.com/abstract=3593946.

Bratspies, Rebecca M. "Renewable Rikers: A Plan for Restorative Environmental Justice." Available at SSRN, July 24, 2020. https://ssrn.com/abstract=3660113 or doi: 10.2139/ssrn.3660113.

Bratspies, Rebecca M. "Shutting down Poletti: Human Rights Lessons from Environmental Victories." *Papers.ssrn.com.* Rochester, NY. December 28, 2018. https://ssrn.com/abstract=3307512.

Brean, Henry. "MGM Resorts Will Use Solar Arrays to Power Las Vegas Casinos." *Las Vegas Review-Journal*, April 18, 2018. https://www.reviewjournal.com/business/casinos-gaming/mgm-resorts-will-use-solar-array-to-power-las-vegas-casinos/.

Brennan, Shane. "Visionary Infrastructure: Community Solar Streetlights in Highland Park." *Journal of Visual Culture* 16, no. 2 (2017): 167–189. doi: 10.1177/1470412916685743.

Broom, Douglas. "Reef Cubes: Could These Plastic-free Blocks Help Save the Ocean?" *World Economic Forum*, July 3, 2020. https://www.weforum.org/agenda/2020/07/reef-cubes-arc-marine-biodiversity-wind-farms/.

Burger, Andrew. "Microgrids Restore Water, Wastewater Services for Isolated Puerto Rico Communities." *MicrogridKnowledge.com*, April 2, 2019. https://microgrid-knowledge.com/microgrids-water-puerto-rico/.

Burgy, Robert S., "A Dramatistic Analysis of Nevada's Controversy over Solar Net Metering Incentive Policies." UNLV Theses, Dissertations, Professional Papers, and Capstones, 2018: 3225.

C40 Cities Report. "CDP Cities 2013: Summary Report on 110 Global Cities." *C40 Cities Climate Leadership Group, C40.org.* CDP. 2013. https://www.c40.org/researches/c40-cdp-2013-summary-report.

California Museum. "Biography of Henry J. Kaiser." *Californiamuseum.org*, accessed on January 2, 2021. https://www.californiamuseum.org/inductee/henry-j-kaiser.

Calvert, A. M., C. A. Bishop, R. D. Elliot, E. A. Krebs, T. M. Kydd, C. S. Machtans, and G. J. Robertson. 2013. "A Synthesis of Human-related Avian Mortality in Canada." *Avian Conservation and Ecology* 8(2): 11. doi: 10.5751/ACE-00581-080211.

Chapman, Simon. "Wind farms are Hardly the Bird Slayers They're Made Out to Be. Here's Why." *The Conversation*, June 16, 2017. https://theconversation.com/wind-farms-are-hardly-the-bird-slayers-theyre-made-out-to-be-heres-why-79567.

Chow, Lorraine. "Kansas Town Decimated by Tornado Now Runs on 100% Renewable Energy, Should Be Model for Frack-Happy State." *EcoWatch.com.* October 18, 2016. https://www.ecowatch.com/greensburg-kansas-renewable-energy-2051037395.html.

City of Las Vegas. "Sustainability." *Lasvegasnevada.gov,* n.d. Accessed June 30, 2020. https://www.lasvegasnevada.gov/Government/Initiatives/Sustainability.

City of Las Vegas, *Sustainability Reports*, Accessed on January 1, 2021.

Clark, Jeremy. "Emerging Best Practices for Utility Grid Hardening." *Utility Dive.* November 5, 2018.

Climate Central. "American Warming: The Fastest-Warming Cities and States in the U.S." *ClimateCentral.org.* April 17, 2019. https://www.climatecentral.org/news/report-american-warming-us-heats-up-earth-day.

Costantini, Fabrizio. "Cities' Cost Cuttings Leave Residents in the Dark." *New York Times,* December 29, 2011. https://www.nytimes.com/2011/12/30/us/cities-cost-cuttings-leave-residents-in-the-dark.html.

Constantinides, Costa. "How to Power a City." Interview by Melanie La Rosa. June 2017.

Constantinides, Costa. "How to Power a City." Interview by Melanie La Rosa. December 2018.

Cwiek, Sarah. "Highland Park Residents Lighting up Their Streets with Solar Power." *Michigan Public Radio*, The Environment Report. February 21, 2017. https://www.michiganradio.org/post/highland-park-residents-lighting-their-streets-solar-power.

Cwiek, Sarah, and Steve Carmody. "Highland Park Residents Lighting Up Their Streets with Solar Power." *Michiganradio.Org.* 2017. https://www.michiganradio.org/post/highland-park-residents-lighting-their-streets-solar-power.

Daily Detroit Staff. "This City Had Their Streetlights Repossessed. Now Some Are Getting Light Thanks to Solar Power." *Daily Detroit.* 2017. http://www.dailydetroit.com/2017/04/25/city-streetlights-repossessed-now-getting-light-thanks-solar-power/.

Davies, Lincoln L., and Sanya Carley. "Emerging Shadows in National Solar Policy? Nevada's Net Metering Transition in Context." *The Electricity Journal* 30, no. 1 (2017): 33–42.

de Onís, Catalina M. "Energy Colonialism Powers the Ongoing Unnatural Disaster in Puerto Rico." *Frontiers in Communication* 3, no. 2 (January 29, 2018). doi: 10.3389/fcomm.2018.00002.

de Onis, Kathleen. *Spreading Toxicity: Illegal Coal Ash Disposal Practices in The Caribbean.* Indiana University. Ebook, 2015. https://theieca.org/sites/default/files/conference-presentations/coce_2015_boulder/de_onis_kathleen_-612255293.pdf.

Dees, J. Gregory. "The Meaning of Social Entrepreneurship." 1998. http://www.redalmarza.cl/ing/pdf/TheMeaningofsocialEntrepreneurship.pdf.

DeFrancia, Kelsie. "Moving towards 100% Renewable Energy: Drivers behind City Policies and Pledges." State of the Planet. Columbia Climate School. April 24, 2018. https://news.climate.columbia.edu/2018/04/24/moving-towards-100-renewable-energy-drivers-behind-city-policies-pledges/.

Department of Public Service, State of Vermont. "Vermont Electric Utilities." *Vermont Official State Website.* https://publicservice.vermont.gov/electric.

Detroit Historical Society. "Ford Highland Park Plant | Detroit Historical Society." *Detroithistorical.Org.* 2021. https://detroithistorical.org/learn/encyclopedia-of-detroit/ford-highland-park-plant.

DeWaard, J., J.E. Johnson, and S.D. Whitaker. "Out-migration from and Return Migration to Puerto Rico After Hurricane Maria: Evidence From the Consumer Credit Panel." *Population and Environment* 42 (2020): 28–42. doi: 10.1007/s11111-020-00339-5.

Dirul, Ali. "Inspiring Possibilities Through the Power of Solar." *TEDxDetroit*, TEDx Youtube Channel. October 17, 2018. https://www.youtube.com/watch?v=jcqt9MgkWbA.

Dominion Energy. "Coast Virginia Offshore Wind: Project Timeline." *Dominion Energy website*. Accessed June 25, 2021. https://coastalvawind.com/about-offshore-wind/timeline.aspx.

Dostis, Robert. "Presentation: House Energy & Technology: Green Mountain Power Corporation" *Vermont legislature website*, January 19, 2017.

DW Akademie. "How Do Offshore Wind Farms Affect Ocean Ecosystems?" *DW.com*, accessed September 20, 2020. https://www.dw.com/en/how-do-offshore-wind-farms-affect-ocean-ecosystems/a-40969339.

Earthjustice. "Groups Argue for 100% Renewable Energy to the Puerto Rico Energy Bureau." *Earthjustice.org.* February 10, 2020. https://earthjustice.org/news/press/2020/100-percent-renewable-energy-governing-board-of-the-puerto-rico-electric-power-authority.

Earthjustice. "Mapping the Coal Ash Contamination." *Earthjustice.org.* October 6, 2020. https://earthjustice.org/features/map-coal-ash-contaminated-sites.

Earthjustice. "Ruth Santiago, Clean Air Ambassador Biography." *Earthjustice.org.* 2019. https://earthjustice.org/50states/2013/ruth-santiago.

Eichler, Alexander. "Highland Park, Michigan Tearing Out Its Streetlights to Cut Costs." *Huffpost.Com.* 2012. https://www.huffpost.com/entry/highland-park-sreetlights_n_1079909.

Englund, Will. "The Grid's Big Looming Problem: Getting Power to Where It's Needed." *Washington Post*, June 29, 2021. https://www.washingtonpost.com/business/2021/06/29/power-grid-problems/.

Eyoko, Stephanie. "Utilities Need to Harden the Grid as They Green It. Consumers Aren't Ready for the Cost." *Utility Dive.* February 26, 2021.

Fast Company. "Green Mountain Power: Most Innovative Company | Fast Company." *Fast Company.* 2020. https://www.fastcompany.com/company/green-mountain-power.

Feeley, Jef, and Mark Chediak. "Power Company AES Settles Claims That It Killed or Deformed Babies with Dumped Coal Ash." *Bloomberg News.* April 4, 2016, https://www.bloomberg.com/news/articles/2016-04-04/aes-settles-suit-over-coal-ash-dumping-in-dominican-republic.

Fialka, John. "As Hawaii Aims for 100% Renewable Energy, Other States Watching Closely." *Scientific American*, April 27, 2018. https://www.scientificamerican.com/article/as-hawaii-aims-for-100-renewable-energy-other-states-watching

-closely/#:~:text=When%20it%20comes%20to%20generating,percent%20renewable%20energy%20by%202045.

Field, Ralph, Eddie Laboy, Jorge Capellla, Pedro Robles, Carmen González, and Angel Dieppa. "Jobos Bay Estuarine Profile: a National Estuarine Research Reserve." *Coast.Noaa.gov.* U.S. National Oceanic and Atmospheric Administration, June 2008. https://coast.noaa.gov/data/docs/nerrs/Reserves_JOB_SiteProfile.pdf.

Firestone, Jeremy, Cristina L. Archer, Meryl P. Gardner, John A. Madsen, Ajay K. Prasad, and Dana E. Verone. "Opinion: The Time Has Come for Offshore Wind Power in the United States." *Proceedings of the National Academy of Sciences of the United States of America*, published online September 29, 2015. doi: 10.1073/pnas.1515376112.

Flores, Antonio, Krogstad, Jens Manuel. "Puerto Rico's Population Declined Sharply after Hurricanes Maria and Irma." *Pew Research Center.* July 26, 2019.

Florida, Richard. "The Real Powerhouses That Drive the World's Economy." *Citylab/Bloomberg.com.* February 28, 2019. https://www.bloomberg.com/news/articles/2019-02-28/mapping-the-mega-regions-powering-the-world-s-economy.

Foehringer, Emma. "Grid Defection Is On the Rise in Puerto Rico." *GreenTech Media,* February 16, 2018. https://www.greentechmedia.com/articles/read/grid-defection-on-the-rise-in-puerto-rico.

Fournier, Nicquel Terry, and Holly. "Outages Drag Past a Week for Some DTE Customers." *The Detroit News.* March 15, 2015. https://www.detroitnews.com/story/news/local/oakland-county/2017/03/15/detroit-edison-power-outage-stragglers/99230388/.

Francis, Jennifer. "Yes, Climate Change Is Making Severe Weather Worse." *Scientific American,* June 1, 2019. https://www.scientificamerican.com/article/yes-climate-change-is-making-severe-weather-worse/.

Free, Christopher M., James T. Thorson, Malin L. Pinsky, Kiva L. Oken, John Wiedenmann, and Olaf P. Jensen. "Impacts of Historical Warming on Marine Fisheries Production." *Science,* 363, no. 6430 (March 1, 2019): 979–983. doi: 10.1126/science.aau1758.

French, Ron. "Study Claims Michigan Has More Blackouts Than Most States. Not True, Says DTE." *Bridgemi.Com.* 2017. https://www.bridgemi.com/michigan-government/study-claims-michigan-has-more-blackouts-most-states-not-true-says-dte.

Funes, Yessenia. "How One Small Tribe Beat Coal and Built a Solar Plant." *Colorlines.Com.* 2017. https://www.colorlines.com/articles/how-one-small-tribe-beat-coal-and-built-solar-plant.

Gallucci, Maria. "The Privatization of Puerto Rico's Power Grid Is Mired in Controversy." *IEEE Spectrum.* July 8, 2020. https://spectrum.ieee.org/energywise/energy/policy/the-privatization-of-puerto-rico-power-grid-mired-in-controversy.

Garrabrants, A. C., D. S. Kosson, R. DeLapp, Peter Kariher, and Susan A. Thorneloe. "Leaching Behavior of "AGREMAX" Collected from a Coal-Fired Power Plant in Puerto Rico." 2012. https://cfpub.epa.gov/si/si_public_record_report.cfm?Lab=NRMRL&dirEntryId=307594.

Gerke, Thomas. "Energy Democracy—Video & Campaign." *CleanTechnica Campaign Paper*. May 14, 2013. https://cleantechnica.com/2013/05/14/energy-democracy-video-campaign.

Gerrard, Michael. Email to author. August 5, 2020.

Gheorghiu, Iulia. "Executive of the Year: Mary Powell, Green Mountain Power." *Utility Dive*. 2019. https://www.utilitydive.com/news/ceo-mary-powell-vermont-green-mountain-power-dive-awards/566247/.

Gigantiello, Anthony. "How to Power a City." Interview by Melanie La Rosa. August, 2017.

Glassman, Brian. "More Puerto Ricans Move to Mainland United States, Poverty Declines." *Census.gov. The U.S. Census Bureau*, September 26, 2019. https://www.census.gov/library/stories/2019/09/puerto-rico-outmigration-increases-poverty-declines.html.

Graham, David. "Is the Jones Act Waiver All Politics?" *The Atlantic*. 2017. https://www.theatlantic.com/politics/archive/2017/09/jones-act-waiver-puerto-rico-trump/541398/.

Greco, JoAnn. "A Shocking Toll: Saving Eagles from the Lethal Hazards of Power Line Electrocution." *National Wildlife*, National Wildlife Foundation. February 1, 2021. https://www.nwf.org/Home/Magazines/National-Wildlife/2021/Feb-Mar/Animals/Eagles-and-Powerlines.

Green Chips Clark County. "NV 2016 Report." http://www.clarkcountynv.gov/comprehensive-planning/eco-county/announcements/Pages/Green-Chips-Southern-Nevada-State-of-Sustainability-Report-2016.aspx.

Green Mountain Power. "Energy Mix." *Green Mountain Power website*, accessed January 2, 2021. https://greenmountainpower.com/energy-mix/.

Green Mountain Power. "Tesla Powerwall." *Green Mountain Power website*, accessed January 2, 2021. https://greenmountainpower.com/rebates-programs/home-energy-storage/powerwall/.

Greentech Media. "How Storage Can Help Get Rid of Peaker Plants?" *Greentech Media*, June 28, 2010. https://www.greentechmedia.com/articles/read/energy-storage-vs-peakers.

Grossman, David. "Clean Coal Explained - What Is Clean Coal?" *Popular Mechanics*, November 13, 2020.

Guzmán, Alexander, and Trujillo, Maria-Andrea. "Social Entrepreneurship–Literature Review (Emprendimiento Social–Revisión de la Literatura) (Spanish)." *Estudios gerenciales* 24, no. 109 (2008): 109–219.

Haag, Matthew. "N.Y.C. Votes to Close Rikers. Now Comes the Hard Part." *New York Times*. October 17, 2019. https://www.nytimes.com/2019/10/17/nyregion/rikers-island-closing-vote.html.

Halstead, Bob. "Nevada Is Winning the War, but Yucca Mountain Is Not Dead." *Las Vegas Sun*, August 23, 2020. https://lasvegassun.com/news/2020/aug/23/nevada-is-winning-the-war-but-yucca-mountain-is-no/.

Harris, Rhonda. "Chapter 5, Selling Outlook. FUD—Fear, Uncertainty, and Doubt: Those Subtle Messages about Competitors." In *The Complete Sales Letter Book*. Armonk: Sharpe Professional. 1998.

Harvey, Abby, Aaron Larson, and Sonal Patel. "History of Power: The Evolution of the Electric Generation Industry." *Power Magazin/powermag.com*, accessed December 29, 2020. Originally published October 2017, updated December 22, 2020. https://www.powermag.com/history-of-power-the-evolution-of-the-electric-generation-industry/.

Hasberg, Kirsten S., and Tegilus, Bob. "Energy Democracy Defined," *This Week in Energy* podcast. August 9, 2013. https://archive.org/details/hpr1310. See also: https://vbn.aau.dk/en/publications/energy-democracy-defined-episode-of-this-week-in-energy-podcast.

Hashan, Mahamudul, M. Farhad Howladar, Labiba Nusrat Jahan, and Pulok Kanti Deb. "Ash Content and Its Relevance with the Coal Grade and Environment in Bangladesh." *International Journal of the Scientific and Engineering Research* 4, no. 4 (2013): 669–676.

Hendrick, Daniel. "Queens' Bad Air Worsens—Sharp Rise in Toxic Emissions from Power Plants." *Queen's Chronicle*, July 10, 2003. https://www.qchron.com/editions/queenswide/queens-bad-air-worsens-sharp-rise-in-toxic-emissions-from/article_c833c111-fa77-5c32-8086-f4b8c2d4e18a.html.

Hernández, Arelis R. "Puerto Ricans Still Waiting on Disaster Funds as Hurricane Maria's Aftermath, Earthquakes Continue to Affect Life on the Island." *Washington Post*, January 9, 2020. https://www.washingtonpost.com/national/puerto-ricans-still-waiting-on-disaster-funds-as-hurricane-marias-aftermath-earthquakes-continue-to-affect-life-on-the-island/2020/01/19/3864fcea-387f-11ea-bb7b-265f4554af6d_story.html.

Hernandez, Dan. "Behind the Bright Lights of Vegas: How the 24-Hour Party City Is Greening up Its Act." *The Guardian*. May 3, 2016. https://www.theguardian.com/sustainable-business/2016/may/03/las-vegas-strip-sustainability-mgm-caesars-water-resource-drought.

Hinojosa, Jennifer, Edwin Meléndez, and K. Severino Pietri. "Population Decline and School Closure in Puerto Rico." *Center for Puerto Rican Studies at Hunter College*, 2019. https://centropr.hunter.cuny.edu/sites/default/files/PDF_Publications/centro_rb2019-01_cor.pdf.

Homeland Security Department. "Waiver of Compliance with Navigation Laws; Hurricanes Harvey and Irma." *Federal Register*, September 14, 2017. https://www.federalregister.gov/documents/2017/09/14/2017-19523/waiver-of-compliance-with-navigation-laws-hurricanes-harvey-and-irma.

Homeland Security Department. "Waiver of Compliance with Navigation Laws; Hurricane Maria." *Federal Register*, October 4, 2017. https://www.federalregister.gov/documents/2017/10/04/2017-21283/waiver-of-compliance-with-navigation-laws-hurricane-maria.

Hvistendahl, Mara. "Coal Ash Is More Radioactive Than Nuclear Waste." *Scientific American*, December 13, 2007.

HydroQuébec. "From The Power Station to Your Home." *HydroQuébec.com website*, accessed January 2, 2021. http://www.hydroquebec.com/learning/transport/parcours.html.

Independent Commission on New York City Criminal Justice and Incarceration Reform. "A More Just New York City." *Independent Commission on New York City Criminal Justice and Incarceration Reform*, April 2017. https://www.morejustnyc.org/reports.

Irizarry, A., B. Colucci, and Efrain O'Neill. "Achievable Renewable Energy Targets for Puerto Rico's Renewable Energy Portfolio Standard." *Puerto Rico Energy Affairs*, 2008. https://bibliotecalegalambiental.files.wordpress.com/2013/12/achievable-renewableenergy-targets-fo-p-r.pdf and https://www.uprm.edu/aret/.

Jacobson, Mark Z., Cristina L. Archer, and Willett Kempton. "Taming Hurricanes with Arrays of Offshore Wind Turbines." *Nature Climate Change* 4, no. 3 (2014), 195–200.

Jain, Monika, Sushma Gupta, Deepika Masand, Gayatri Agnihotri, and Shailendra Jain. "Real-Time Implementation of Islanded Microgrid for Remote Areas." *Journal of Control Science and Engineering*, April 18, 2016. https://www.hindawi.com/journals/jcse/2016/5710950/.

Jones-Albertus, Becca. "Confronting the Duck Curve: How to Address Over-Generation of Solar Energy." *Energy.Gov*, October 12, 2017. https://www.energy.gov/eere/articles/confronting-duck-curve-how-address-over-generation-solar-energy.4.

Katz, Jonathan. "The Disappearing Schools of Puerto Rico." *New York Times*, September 12, 2019. https://www.nytimes.com/interactive/2019/09/12/magazine/puerto-rico-schools-hurricane-maria.html?mtrref=undefined&gwh=5C5ABE1DDFAE6EA03ACE7247929ADC85&gwt=pay&assetType=PAYWALL.

Kaufman, Alexander. "On Puerto Rico's 'Forgotten Island,' Tesla's Busted Solar Panels Tell a Cautionary Tale." *Huffpost.Com*, May 17, 2019. https://www.huffpost.com/entry/elon-musk-tesla-puerto-rico-renewable-energy_n_5ca51e99e4b082d775dfec35.

Kaufman, Alexander. "Power Plant Accident Casts New Light on New York'S Dirty Fuel Addiction." *Huffpost.Com*. 2019. https://www.huffpost.com/entry/transformer-explosion-nyc-fuel-addiction_n_5c25c357e4b0407e9081305a?guccounter=1.

Keegan, James. "Offshore Windmill's Impact on the Marine Environment." *University of Miami, Shark Research blog*, March 4, 2015. https://sharkresearch.rsmas.miami.edu/offshore-windmills-impact-on-the-marine-environment/.

Kelkar, Kamala, Ivette Feliciano, and Zachary Green. "Residents of This City Already Worried about the Coal-Burning Plant Nearby. Then Came Hurricane Maria." *TV News Series Episode*. PBS Newshour, April 28, 2018. https://www.pbs.org/newshour/health/residents-of-this-city-already-worried-about-the-coal-burning-plant-nearby-then-came-hurricane-maria.

Kern, Rebecca. "Rooftop Solar Nearly Doubles in Puerto Rico One Year after Maria." *Bloomberglaw.com. Bloomberg Law*, September 20, 2018. https://news.bloomberglaw.com/environment-and-energy/rooftop-solar-nearly-doubles-in-puerto-rico-one-year-after-maria.

Klein, Naomi. *The Shock Doctrine: The Rise of Disaster Capitalism*. New York: Picador, 2007.

Kravchenko, Julia, Lyerly, H. Kim. "The Impact of Coal-powered Electrical Plants and Coal Ash Impoundments on the Health of Residential Communities." *North Carolina Medical Journal* 79, no. 5 (2018): 289–300.

KTNV Channel 13 Las Vegas. "New Solar Plant Near Boulder City." *KTNV Channel, 13 Las Vegas Youtube channel*, February 14, 2017. https://www.youtube.com/watch?v=lZ5N6Ne0TSw&ab_channel=KTNVChannel13LasVegas.

Kummitha, Rama Krishna Reddy. "Social Entrepreneurship, Energy and Urban Innovations." In *Mainstreaming Climate Co-Benefits in SIndian Cities*, pp. 265–283. Singapore: Springer, 2018.

La Fundación Segarra Boerman e Hijos, Inc. "Fundación Segarra Boerman E Hijos, Inc." *Fsbpr.org*. *Fundación Segarra Boerman e Hijos, Inc.* Accessed June 17, 2020. https://www.fsbpr.org.

Laitner, Bill. "Tiny Michigan Nonprofit is Taking on DTE—And It Could Have a Huge Impact on the Company." *Detroit Free Press*. January 16, 2020. https://www.freep.com/story/news/local/michigan/wayne/2020/01/16/dte-soulardarity-energy-democracy/4429251002/.

LaRosa Field Notes, 2018. https://www.yarotek.com/yarotek-commits-to-local-communities-after-irma-and-maria/.

Lee, Paul. "Highland Park Streetlight Removal." Paul Lee YouTube Channel, 2011.

Legislature of Puerto Rico. "Act No. 17-2019." *Autoridad de Energia Eléctrica, AEEPR .com*. April 11, 2019.

Lehman, Robert N., Patricia L. Kennedy, and Julie A. Savidge. "The State of the Art in Raptor Electrocution Research: A Global Review." *Biological Conservation* 136, no. 2 (April 2007): 159–174. doi: 10.1016/j.biocon.2006.09.015.

Leichenko, Robin M., Melanie Hughes Mcdermott, Ekaterina Bezborodko, Michael Brady, and Erik Namendorf. "Economic Vulnerability to Climate Change in Coastal New Jersey: a Stakeholder-Based Assessment." *Journal of Extreme Events* (June 2014). doi: 10.1142/S2345737614500031.

Lewis, Andrew S. "Why Atlantic City's Minority Neighborhoods Are Also its Most Flooded." *NJ Spotlight News*. April 5, 2021. https://www.njspotlight.com/2021/04/redlining-atlantic-city-nj-overlooked-underfunded-minority-neighborhoods-back-bay-racist-maps-superstorm-sandy/.

Lloréns-Vélez, Eva. "Puerto Rico Senator Rails Against Utility's Restructuring Support Agreement." *Caribbean Business*. May 14, 2019. https://caribbeanbusiness.com/puerto-rico-senator-rails-against-utilitys-restructuring-support-agreement/?cn-reloaded=1.

Luskin Center for Innovation. *Progress Toward 100% Clean Energy in Cities and States Across the U.S.* UCLA Luskin Center for Innovation (LCI). November 6, 2019. https://innovation.luskin.ucla.edu/wp-content/uploads/2019/11/100-Clean-Energy-Progress-Report-UCLA-2.pdf.

Lyubich, Eva. "The Race Gap in Residential Energy Expenditures." 2020. https://haas.berkeley.edu/wp-content/uploads/WP306.pdf.

Maher, Kris. "Trump Administration Revives Nevada Plan as Nuclear Waste Piles Up." *Wall Street Journal*, May 9, 2017. https://www.wsj.com/articles/a-glimmer-of-hope-in-a-growing-hoard-of-nuclear-waste-1494345622.

Mapp, Kenneth. "Written Testimony of Governor Kenneth E. Mapp of the United States Virgin Islands before the Senate Committee on Energy and Natural Resources." 2017. https://www.energy.senate.gov/services/files/A2538A49-2953-4BA1-8C94-0807E62050A5.

Marcel, Joyce. "She was Fast, Fun and Effective: Mary Powell leaves GMP." *Vermont Business Magazine*. January 18, 2020. https://vermontbiz.com/news/2020/january/18/she-was-fast-fun-and-effective-mary-powell-leaves-gmp.

Maria Fund. "Iniciativa de Ecodesarrollo de Bahia de Jobos." *MaríaFund.org*. n.d. Accessed June 30, 2020. https://www.mariafund.org/idebajo.

Marsooli, Reza, and Lin, Ning. "Impacts of Climate Change on Hurricane Flood Hazards in Jamaica Bay, New York" *Climatic Change* (November 26, 2020). doi: 10.1007/s10584-020-02932-x.

Martinez, Hayley, DeFrancia, Kelsie, and Schroder, Alix. "Moving Towards 100% Renewable Energy: Drivers Behind City Policies and Pledges." Synopsis published on *State of The Planet, Earth Institute*, April 24, 2018. Full report presented at the 76th Annual Midwest Political Science Association Conference, 2018.

Massol-Deyá, Arturo, Jennie C. Stephens, and Jorge L. Colón. "Renewable Energy for Puerto Rico." *Science*, 362, no. 6410 (October 5, 2018): 7. doi: 10.1126/science.aav5576.

May 2017, Hilda Lloréns / 5. 2017. "The Making of a Community Activist." SAPIENS. May 5, 2017. https://www.sapiens.org/culture/jobos-bay-community-activist.

McKinley, Jesse, and Brad Plumer. "New York to Approve One of the World's Most Ambitious Climate Plans." *New York Times*, June 18, 2019. https://www.nytimes.com/2019/06/18/nyregion/greenhouse-gases-ny.html.

Metro New York. "No Easy Answers on Astoria Borealis." *Metro.us*, February 11, 2019. https://www.metro.us/no-easy-answers-in-astoria-borealis-transformer-fire-hearing/.

Molony, Ray. "Major Study Finds Outdoor Lighting Cut Crime by 39%." *LEDs Magazine*, Originally published on *Luxreview.com*, March 1, 2018. Republished through partnership, April 20, 2020 https://www.ledsmagazine.com/architectural-lighting/article/16701511/major-study-finds-outdoor-lighting-cut-crime-by-39.

"More on That Metaphor About the Electric Grid as a Bathtub." *Concord Monitor*, November 14, 2017. https://www.concordmonitor.com/electricity-power-grid-13516204.

Morehouse, Catherine. "Nevada Passes Bill for 50% Renewables by 2030, 100% Carbon Free by 2050." *Utility Dive,* April 22, 2109.

NASA. "NASA Satellite Confirms Sharp Decline in Pollution from US Coal Power Plants," NASA Goddard Science Center. December 01, 2011.

National Audubon Society. "Wind Power and Birds: Properly Sited Wind Power Can Help Protect Birds from Climate Change." *National Audubon Society website*, July 21, 2020. https://www.audubon.org/news/wind-power-and-birds.

National Oceanic and Atmospheric Administration, Ranking of Cities Based on Percentage of Annual Possible Sunshine. https://www1.ncdc.noaa.gov/pub/data/ccd-data/pctposrank.txt.

National Oceanic and Atmospheric Administration. "The State of High Tide Flooding and Annual Outlook." *Tides & Currents website.* Accessed July 1, 2021. https://tidesandcurrents.noaa.gov/HighTideFlooding_AnnualOutlook.html.

National Renewable Energy Lab. "A Guide to Community Shared Solar: Utility, Private and Non-profit Project Development." *National Renewable Energy Lab.* U.S. Department of Energy. May 2012.

Nevada Commission on Nuclear Projects. "Report and Recommendations, November 2019." www.State.NV.us/Nucwaste/. *State of Nevada.* November 2019. Website accessed on January 2, 2021. http://www.state.nv.us/nucwaste/pdf/2019%20Commission%20Report%20and%20Recommendations.pdf.

New Jersey Conservation Foundation. "Fish in Hot Water." *NJConservation.org,* July 3, 2019. https://www.njconservation.org/fish-in-hot-water/.

New Jersey State. "The Regional Greenhouse Gas Initiative in New Jersey." *Department of Environmental Protection,* accessed December 29, 2020. https://www.state.nj.us/dep/aqes/rggi.html.

New Jersey State Legislature. "The Offshore Wind Economic Development Act (S-2036), otherwise known as the Sweeney/Kean Bill." *NJsendems.org,* accessed December 29, 2020. https://www.njsendems.org/sweeneykean-bill-to-spur-offshore-wind-energy-released-by-senate-committee.

New York City. "Benchmarking and Energy Efficiency Grading." *NYC.gov,* accessed December 29, 2020. https://www1.nyc.gov/site/buildings/business/benchmarking.page.

New York City Energy Policy Task Force. "New York City Energy Policy: An Electricity Resource Roadmap." *New York City Energy Policy Task Force,* January 2004. http://www.nyc.gov/html/om/pdf/energy_task_force.pdf.

New York City League of Conservation Voters. "New York City Climate Action Tracker." *Climatetracker.nylcvef.org,* accessed December 29, 2020. https://climatetracker.nylcvef.org/.

New York ISO. "Gold Book - 2019 Load and Capacity Data." *New York Independent System Operator.* 2019. https://www.nyiso.com/documents/20142/2226333/2019-Gold-Book-Final-Public.pdf/a3e8d99f-7164-2b24-e81d-b2c245f67904.

New York ISO. "Power Trends 2021: New York's Clean Energy Grid of the Future." *New York Independent System Operator Annual Grid and Markets Report.* 2021. https://www.nyiso.com/documents/20142/2223020/2021-Power-Trends-Report.pdf/471a65f8-4f3a-59f9-4f8c-3d9f2754d7de.

New York Power Authority, Puerto Rico Electric Power Authority, Puerto Rico Energy Commission, Consolidated Edison, Edison International, Electric Power Research Institute, Long Island Power Authority, et al. 2017. "Build Back Better: Reimagining and Strengthening the Power Grid of Puerto Rico." *Governor.ny.gov.* New York State. https://www.governor.ny.gov/sites/default/files/atoms/files/PRERWG_Report_PR_Grid_Resiliency_Report.pdf#:~:text=On%20behalf%20of%20the%20Working.

New York State Independent System Operator. "What We Do." *NYISO.com,* accessed December 29, 2020. https://www.nyiso.com/what-we-do.

New York State Reliability Council. "About New York State Reliability Council." *Nysrc.org*, accessed December 29, 2020. http://www.nysrc.org/.

Newbery, Charles. "Energy Storage Poses a Growing Threat to Peaker Plants." *General Electric, Transform blog*, October 1, 2018. https://www.ge.com/power/transform/article.transform.articles.2018.oct.storage-threat-to-peaker-plants.

Nichols, Shimekia. "How to Power a City." Interview by Melanie La Rosa. September, 2017.

NOAA Fisheries Public Affairs. "New Study: Climate Change to Shift Many Fish Species North." *NOAA*. Last modified 2019. https://www.fisheries.noaa.gov/feature-story/new-study-climate-change-shift-many-fish-species-north.

NOAA Office for Coastal Management. "Jobos Bay National Estuarine Research Reserve." *National Oceanic and Atmospheric Administration website*. Accessed June 23, 2021. https://coast.noaa.gov/data/docs/nerrs/Handout-Jobos-Bay.pdf.

NV Energy. "Nellis Solar Array II Generating Station Flyer." Website accessed January 2, 2021. https://www.nvenergy.com/publish/content/dam/nvenergy/brochures_arch/about-nvenergy/our-company/power-supply/Nellis-Fact-Sheet.pdf.

Obama, Barack. "A Shining Light - Summary: Walter Meyer Is Being Honored as a Champion of Change for the Leadership He Demonstrated in His Involvement in Response and Recovery Efforts Following Hurricane Sandy." *The White House blog*, June 19, 2013. https://obamawhitehouse.archives.gov/blog/2013/06/19/shining-light.

Oceana. "Offshore Wind Report: Key Findings." *USA.Oceana.Org*. Accessed January 2. https://usa.oceana.org/offshore-wind-report-key-findings.

Pérez-Peña, Richard. "State to Close Queens Plant That Is Biggest Polluter in City (Published 2002)." *New York Times*, September 5, 2002. https://www.nytimes.com/2002/09/05/nyregion/state-to-close-queens-plant-that-is-biggest-polluter-in-city.html.

Perkins, Tom. "Michigan's Energy Activists Start to See a Decade of Work Pay Off in Highland Park." *Energy News Network*. May 4, 2021. https://energynews.us/2021/05/04/michigans-energy-activists-start-to-see-a-decade-of-work-pay-off-in-highland-park/.

Pothecary, Sam. "The Largest Solar Rooftop System in the US Installed on the Mandalay Bay Convention Center." *PV Magazine*, July 7, 2016.

Powell, Mary. "How to Power a City." Interview by Melanie La Rosa. October, 2018.

Puerto Rico Energy Bureau. "Testimony of Professor Agustín Irizarry-Rivera, Page 9, definition of "prosumer." *Energia.pr.gov*. 2019. https://energia.pr.gov/wp-content/uploads/2019/10/LEOs-Motion-for-Submission-of-Testimony-with-Testimonies.pdf.

Pushpendra Kumar Singh Rathore, Shyam Sunder Das, and Durg Singh Chauhan. "Perspectives of Solar Photovoltaic Water Pumping for Irrigation in India." *Energy Strategy Reviews*, 22 (2018): 385–395. doi: 10.1016/j.esr.2018.10.00.

Queremos Sol (We Want Solar). "Queremos Sol Propuesta, Versión 4.0 (We Want Solar Proposal v. 4.0)." *Queremos Sol*. February 1, 2020. https://www.queremossolpr.com/.

Renewables Now. "EDF RE to acquire Fishermen's 24-MW wind demo off Atlantic City." *RenewablesNow.com*, April 5, 2018. https://renewablesnow.com/news/edf-re-to-acquire-fishermens-24-mw-wind-demo-off-atlantic-city-607792/.

Retail Energy Supply Association. "Energy Glossary." *Resausa.org*, accessed December 29, 2020. https://www.resausa.org/shop-energy/energy-glossary#r.

Richardson, Jake. "Wind Power Results in Very Few Bird Deaths Overall." *Clean Technica*, February 21, 2018. https://cleantechnica.com/2018/02/21/wind-power-results-bird-deaths-overall/.

Ritter, Ken. "NV Energy Pulls Plug on Coal-fired Power Plant Near Las Vegas." *Las Vegas Sun*, March 16, 2017. https://www.lasvegasnevada.gov/Government/Initiatives/Sustainability/Sustainability-Resources?tab=0.

Roblin, Stéphanie. Solar-powered Irrigation: A Solution to Water Management in Agriculture?" *Renewable Energy Focus*, 17, no. 5 (2016): 205–206, doi: 10.1016/j.ref.2016.08.013.

Roff, Kate. "COVID-19 Shines a Light on Accessible Solar Power's Role in Detroit Communities." *Model-D Media*. December 15, 2020.

Rosa-Aquino, Paola. "What Will a Jones Act Waiver Mean for Puerto Rico's 100% Renewable Energy Goal?" *Grist*, April 26, 2019. https://grist.org/article/what-will-a-jones-act-waiver-mean-for-puerto-ricos-100-renewable-energy-goal/.

Roselló, Ricardo. "The Need For Transparent Financial Accountability in Territories' Disaster Recovery Efforts." *Congressional Testimony*, Committee on Natural Resources, November 14, 2017. https://www.govinfo.gov/content/pkg/CHRG-115hhrg27587/html/CHRG-115hhrg27587.htm.

Roston, Eric and Wade, Will. "As Indian Point Goes Dark, New York Races to Swap Nuclear With Wind." *Bloomberg Green, Energy & Science*. April 30, 2021. https://www.bloomberg.com/news/articles/2021-04-30/indian-point-nuclear-plant-shuts-down-and-new-york-races-for-wind-power.

Sabin Center for Climate Change Law. "New York City Climate Law Tracker." *Climate.Law.Columbia.Edu*. Sabin Center for Climate Change Law, Columbia Climate School. Accessed June 15, 2021. https://climate.law.columbia.edu/content/nyc-climate-law-tracker.

Santiago, Ruth, de Onís, Catalina M., Cataldo, Kenji, and Lloréns, Hilda. "A Disastrous Methane Gas Scheme Threatens Puerto Rico's Energy Future." *Nacla.org*, June 4, 2020. https://nacla.org/news/2020/06/04/methane-gas-scheme-puerto-rico-energy.

Santiago, Ruth. "How to Power a City." Interview by Melanie La Rosa. March, 2018.

Santiago, Ruth. "How to Power a City." Interview by Melanie La Rosa. December, 2018.

Santiago, Ruth. "Imminent and Substantial Endangerment to Human Health and the Environment from Use of Coal Ash as Fill Material at Construction Sites in Puerto Rico: A Case Study." *Procedia - Social and Behavioral Sciences* 37, no. (January 1, 2012): 389–396. doi: 10.1016/j.sbspro.2012.03.304.

Sawle, Yashwant, S.C. Gupta, and Aashish Kumar Bohre. "Review of Hybrid Renewable Energy Systems with Comparative Analysis of Off-grid Hybrid

System." *Renewable and Sustainable Energy Reviews*, 81, no. 2 (2018): 2217–2235. doi: 10.1016/j.rser.2017.06.033.

Schlossberg, Tatiana. "America's First Offshore Wind Farm Spins to Life." *New York Times*, December 14, 2016. https://www.nytimes.com/2016/12/14/science/wind-power-block-island.html.

Shahan, Zachary. "Top US States for Percentage of Electricity from Solar." *CleanTechnica*. October 5, 2020. https://cleantechnica.com/2020/10/05/top-us-states-for-percentage-of-electricity-from-solar-cleantechnica-report/.

Shahrigian, Shant. "ConEd Grid Guys Grilled by City Council For 'Inadequate and Laughable' Response to Summer Blackout." *New York Daily News*. September 4, 2019. https://www.nydailynews.com/news/politics/ny-coned-city-council-corey-johnson-grilling-20190904-v7crvydvc5ebdpe43rcj7xki4m-story.html.

Shannon, Juan. "How to Power a City." Interview by Melanie La Rosa. May 2018.

Shannon, Juan. "How to Power a City." Interview by Melanie La Rosa. November, 2018.

Shaw, Eleanor, and Sara Carter. "Social Entrepreneurship: Theoretical Antecedents and Empirical Analysis of Entrepreneurial Processes and Outcomes." *Journal of Small Business and Enterprise Development* 14, no. 3 (2007): 418–434. doi: 10.1108/14626000710773529.

Shogran, Elizabeth. "In Solar Scuffle, Big Utilities Meet their Match." *High Country News*, August 21, 2017. https://www.hcn.org/issues/49.14/solar-energy-solar-eclipse-big-utilities-meet-their-match-in-solar-scuffle.

Siemens Energy. "Combined Heat and Power Brochure." *Siemens-Energy.com* (Global Website). Siemens, n.d. Accessed January 22, 2020. https://www.siemens-energy.com/global/en/offerings/power-generation/power-plants/combined-heat-and-power.html.

Solar Energy Industries Association. "Community Solar." *Solar Energy Industries Association*. SEIA. n.d. Accessed June 22, 2020. https://www.seia.org/initiatives/community-solar.

Solar Energy Industries Association. "Residential Consumer Guide to Solar Power." *Solar Energy Industries Association Website*. SEIA, 2018. Website accessed June 30, 2020. https://www.seia.org/sites/default/files/2018-06/SEIA-Consumer-Guide-Solar-Power-v4-2018-June.pdf.

Solar Energy Industries Association. "Solar Industry Research Data | SEIA." SEIA. Website accessed January 22, 2021. https://www.seia.org/solar-industry-research-data.

Solar Energy Industries Association. "Solar Industry Urges U.S. District Court to Reject Discriminatory Charges on Puerto Rico Solar Customers." Press Release, *Solar Energy Industries Association*. October 25, 2019. https://www.seia.org/news/solar-industry-urges-us-district-court-reject-discriminatory-charges-puerto-rico-solar.

Sonal, Jessel; Sawyer, Samantha, and Hernández, Diana. "Energy, Poverty, and Health in Climate Change: A Comprehensive Review of an Emerging Literature." *Frontiers in Public Health,* 7 (December 12, 2019): 357. doi: 10.3389/fpubh.2019.00357.

Soulardarity Press Release. "Michigan Nonprofit Soulardarity Names New Executive Director." *AlternativeEnergyMag.com* and *Soulardarity.com.* April 22. 2021.

Soulardarity website. "Why Energy Democracy." *Soulardarity.com.* Accessed June 1, 2021. https://www.soulardarity.com/why_energy_democracy.

Soulardarity.com. "2019 Annual Report." *Soulardarity.com.* Accessed June 15, 2021.

Sovacool, Benjamin K. "Contextualizing Avian Mortality: A Preliminary Appraisal of Bird and Bat Fatalities from Wind, Fossil-fuel, and Nuclear Electricity," *Energy Policy*, 37, no. 6 (2009): 2241–2248.

Sovacool, Benjamin, Alex Gilbert, and Brian Thomson. 2014a. "Innovations in Energy and Climate Policy: Lessons from Vermont." *Pace Environmental Law Review* 31(3): 651.

Spear, Kevin. "Osceola County landfill takes in coal ash from Puerto Rico, triggering public backlash." *Orlando Sentinel*, May 13, 2019.

Spector, Julian. "How MGM Prepared Itself to Leave Nevada's Biggest Utility: The Casino Conglomerate Expects to Double Its Use of Renewable Energy and Earn Payback Within 7 Years." *Greentech Media*, September 16, 2016.

Stephens, Jennie C. "Energy Democracy: Redistributing Power to the People Through Renewable Transformation." *Environment: Science and Policy for Sustainable Development*, 61, no. 2 (February 13, 2019): 4–13, doi: 10.1080/00139157.2019.1564212.

Stevens, Matt, Rick Rojas, and Jacey Fortin. "New York Sky Turns Bright Blue After Transformer Explosion." *New York Times*, December 28, 2019. https://www.nytimes.com/2018/12/27/nyregion/blue-sky-queens-explosion.html.

Stone, Laurie, and Burgess, Christopher. "Solar Under Storm: Designing Hurricane-Resilient PV Systems." *Rocky Mountain Institute website*. June 20, 2018. https://rmi.org/solar-under-storm-designing-hurricane-resilient-pv-systems/.

Surrusco, Emilie Karrick. "In the Fight to Clean up Coal Ash, These States Are Making Progress." *EarthJustice*, February 6, 2020. https://earthjustice.org/blog/2020-february/in-the-fight-to-clean-up-coal-ash-these-states-are-making-progress.

Swain, Marian, Hsu, David, and Bui, Lily. "Developing a Resilient Energy Infrastructure for Puerto Rico." *MIT, Department of Urban Planning website*. 2018. https://dusp.mit.edu/epp/news/developing-resilient-energy-infrastructure-puerto-rico.

T&D World Magazine. "Green Mountain Power Files Plans to Offer Tesla Powerwalls to Customers." *T&D World Magazine website*, December 8, 2015. https://www.tdworld.com/distributed-energy-resources/article/20965959/green-mountain-power-files-plans-to-offer-tesla-powerwalls-to-customers.

Tesla, "Exclusive Green Mountain Power Solar and Powerwall Offer." *Tesla.com*, accessed January 2, 2021. https://www.tesla.com/gmp-bundle.

The Avalon Village. "The Avalon Village." *The Avalon Village official website*, accessed December 23, 2020. http://theavalonvillage.org/.

Toplikar, Dave. "New Las Vegas City Hall dedication a highlight of downtown development." *Las Vegas Sun*, Monday, March 5, 2012 https://lasvegassun.com/news/2012/mar/05/new-city-hall-dedicated-light-show/.

Torbert, Roy, and Mike Henchen. "Implementing Puerto Rico's Energy Transformation." *Rocky Mountain Institute*, March 10, 2020. https://rmi.org/implementing-puerto-ricos-energy-transformation/.

U.S. Census. "U.S. Census Bureau QuickFacts: Las Vegas City, Nevada." *Census.gov*. July 1, 2019. https://www.census.gov/quickfacts/lasvegascitynevada.

U.S. Census Bureau. "The Puerto Rico Community Survey Annual Data." *Census.gov: U.S. Census*. Accessed December 23, 2020. https://www.census.gov/programs-surveys/acs/about/puerto-rico-community-survey.html.

U.S. Conference of Mayors, "Mayors Climate Protection Agreement." Resolution passed in 2005. Accessed January 2, 2021. https://wwwusmayors.org/programs/mayors-climate-protection-center/.

U.S. Department of Energy. "Confronting the Duck Curve: How to Address Over-Generation of Solar Energy." *Office of Energy Efficiency & Renewable Energy*, October 12, 2017.

U.S. Department of Energy. "DOI Announces Two Solar Projects Approved in California, Nevada." *Department of Energy website*, February 26, 2014. https://www.energy.gov/eere/solar/articles/doi-announces-two-solar-projects-approved-california-nevada.

U.S. Department of Energy. "Grid-Connected Renewable Energy Systems." *Energy.gov*, accessed December 29, 2020. https://www.energy.gov/energysaver/grid-connected-renewable-energy-systems.

U.S. Department of Energy. "Learn More About Interconnections." *Energy.gov*, accessed December 29, 2020. https://www.energy.gov/oe/services/electricity-policy-coordination-and-implementation/transmission-planning/recovery-act-0.

U.S. Department of Energy. "Wildlife Impacts of Wind Energy." U.S. Department of Energy, Wind Energy Technologies Office. Accessed May 15, 2021. https://windexchange.energy.gov/projects/wildlife#:~:text=Research%20shows%20that%20wind%20projects%20actually%20rank%20near,posed%20to%20birds%20and%20people%20by%20climate%20change.

U.S. Energy Information Administration. "Indian Point, Closest Nuclear Plant to New York City, Set to Retire by 2021." *Eia.gov*, February 1, 2017. https://www.eia.gov/todayinenergy/detail.php?id=29772.

U.S. Energy Information Administration. "Puerto Rico Profile, Territory Profile and Energy Estimates." Accessed July 2, 2020. https://www.eia.gov/state/analysis.php?sid=RQ.

U.S. Energy Information Administration. "Puerto Rico's Electricity Service is Slow to Return after Hurricane Maria." U.S. EIA. October 2017.

U.S. Energy Information Administration. "Solar Thermal Power Plants." *EIA.gov*, website accessed November 2, 2020. https://www.eia.gov/energyexplained/solar/solar-thermal-power-plants.php.

U.S. Energy Information Administration. "Today in Energy: New York's Indian Point Nuclear Power Plant Closes After 59 Years of Operation." *Eia.gov*. April 30, 2021. https://www.eia.gov/todayinenergy/detail.php?id=47776.

U.S. Energy Information Administration, and Hoff, Sara (Principal Contributor). "EIA Electricity Sales Data for Puerto Rico Show Rate of Recovery since

Hurricanes - Today in Energy - U.S. Energy Information Administration (EIA)." *Eia.gov*. Energy Administration Association. August 6, 2018. https://www.eia.gov/todayinenergy/detail.php?id=36832.

U.S. Environmental Protection Agency. "Distributed Generation of Electricity and its Environmental Impacts." *Epa.gov*, accessed December 29, 2020. https://www.epa.gov/energy/distributed-generation-electricity-and-its-environmental-impacts.

U.S. Environmental Protection Agency. "Re-Powering America's Land: Siting Renewable Energy on Potentially Contaminated Land and Mine Sites, Nellis Air Force Base, Nevada Success Story." February 2009. https://www.epa.gov/sites/production/files/2015-04/documents/success_nellis_nv.pdf.

U.S. EPA. "Disposal of Coal Combustion Residuals from Electric Utilities Rulemakings." *EPA website*. https://www.epa.gov/coalash/coal-ash-rule.

UCLA Luskin Center for Innovation. "Progress toward 100% Clean Energy in Cities and States Across the US." *Luskin Center for Innovation, University of California, Los Angeles*. November 2019. https://innovation.luskin.ucla.edu/wp-content/uploads/2019/11/100-Clean-Energy-Progress-Report-UCLA-2.pdf.

United Nations. "Sustainable Development Goals, Goal 6: Ensure Access to Water and Sanitation for All." *UN.org*, accessed December 29, 2020. https://www.un.org/sustainabledevelopment/water-and-sanitation/.

Valley News. "Lowell, Vt., Supports GMP Wind Power Project — Again." *Vnews.com*, March 5, 2015. https://www.vnews.com/Archives/2014/03/a13WireVtTM-ls-vn-030514.

VELCO. "Who's who in Vermont's Electric System." *Vermont Electric Power Company*. https://www.velco.com/about/learning-center/vermonts-electric-system.

Vogel, Ed. "NV Energy Launches Discount LED Lighting Program for Consumers." *Las Vegas Review-Journal*, February 19, 2014. https://www.reviewjournal.com/news/nv-energy-launches-discount-led-lighting-program-for-consumers/.

Wagner, James and Robles, Frances. "Puerto Rico Is Once Again Hit by an Islandwide Blackout." *New York Times*. April 18, 2018. https://www.nytimes.com/2018/04/18/us/puerto-rico-power-outage.html.

Water Mission. *Watermission.org*, accessed December 29, 2020.

WesTech Engineering. "Wastewater Treatment for Power Plants: Considering Zero Liquid Discharge," *WesTech-inc.com*. September 5, 2017.

William, Driscoll. "Puerto Rico Utility Favors LNG over Solar in Siemens Plan." *PV Magazine*, November 14, 2018. https://www.pv-magazine.com/2018/11/14/puerto-rico-utility-favors-lng-over-solar-in-siemens-plan/.

Wong S. "Decentralised, Off-Grid Solar Pump Irrigation Systems in Developing Countries—Are They Pro-poor, Pro-environment and Pro-women?" In Castro, P., Azul, A., Leal Filho, W., and Azeiteiro, U. (eds.), *Climate Change-Resilient Agriculture and Agroforestry. Climate Change Management*. Cham: Springer.

Zhang, Sarah. "The White House Revives a Controversial Plan for Nuclear Waste: Yucca Mountain is Back, and Nevadans are Not Happy." *The Atlantic*, March 21, 2017. https://www.theatlantic.com/science/archive/2017/03/yucca-mountain-trump/519972/.

Index

Act 17 of Puerto Rico, 119–20
Agremax, 110, 111, 113, 114
Aguirre Power Station, 110
American Wind Energy Association (AWEA), 186
Applied Energy Systems (AES) Power Plant, 110–15
"Ask Me About Power" slogan, 171
Astoria, Queens, 58–67; asthma alley, 60; City Council member, 61–66; community-wide campaign, 60–67
Astoria Energy complex, 59, 62; explosion at, 66–67
Atlantic City, NJ, 23–47; birds and wind turbines, 36–43; Fishermen's Energy, 23, 25, 31–36; offshore wind rates, 44–45
Atlantic County Utility Authority (ACUA), 23, 24–25, 27–31
Auffhammer, Maximilian, 166

Baker, Mark, 150, 153
bathtub metaphor, 13–15, 16, 193
batteries/battery technology, 195
B Corporations (B Corps): GMP, 81–82; well-known, 81
birds, death of, 36–43; Audubon Society on, 37–38; by electrocution from power lines, 41–43; by fossil fuel power, 40–41; FUD narrative, 37–41; by nuclear power plants, 42; onshore *vs.* offshore wind farms, 39–40; by wind trubines, 36–41
Black families, 166–67
blackouts, 15, 18, 52, 55, 66, 88, 95, 96, 115, 129, 138, 151, 152, 164, 167–68
Bloomberg Law, 100
Bloomberg News, 112
Blue Planet Energy, 85–86, 150, 151, 153–55, 156
Bolger, Kyle, 150, 154–55
Bolstad, Jennifer, 131, 134–36, 138, 139, 142, 143, 146–47
Boulder City solar plant, 193
Boulder Solar II, 197
Bratspies, Rebecca, 53, 58, 61, 67, 71, 72
"Build It Back Better," 121
built environment, 183
Burdock, Liz, 44
Business Network for Offshore Wind, 25, 43

California, 38
casinos, 198–99
Centro Communitario El Coquí, 97
Christie, Chris, 32, 35
Chrysler, 169

citizen science, 156
Citizens Utility Board, 167
clean air regulations, 111
clean and renewable energy: concept, 1, 19, 20; opposition to, 18; overview, 1–3; utilities, 17
clean coal, 111
climate change, 7, 130
Climate Mobilization Act, 65. *See also* Green New Deal, New York City
Clinton, Hillary, 145–46
Clinton Global Initiative, 134, 138
coal ash, 110–15
coal-burning power plants, 13, 16, 53, 57, 58, 67, 193, 197–98; sulfur dioxide levels from, 111
Coal Combustion Residuals Rule, 113, 114
Cobra, 103
Coffey, Monica, 24
Colchester, Vermont, 77–91. *See also* Green Mountain Power (GMP)
colonialism, 102. *See also* energy coloniality/colonialism
Colucci-Rios, José, 108
Columbia University's Earth Institute, 185
Comite Dialogo Ambiental, Inc., 97, 110
community development, 71, 174
community knowledge, 107
community solar, 8–9; advantages of, 105, 107–8; defined, 19, 104; governing organization for, 106; Michigan, 159, 164, 173; New York City, 8–9, 53, 67–68; off-grid, 106; Puerto Rico, 97–98, 99, 102, 104–6, 107, 109, 113–14
community wind. *See* community solar
ConEdison, 58, 67
conflicts of interest, 120–21
Constantinides, Costa, 53, 59, 61–66, 67, 70–72
Coqui Solar, Puerto Rico, 97–109, 110, 119, 120; emergency power, 115–18;

solar kit, 116–17; training program, 116. *See also* community solar
corporate culture at GMP, 89–91
Costa Sur plant, 121–22
Covid-19, 122
Covid-19 pandemic, 71, 73, 167–68, 177, 178, 205
Cruz, Victor, 143
Culebra, Puerto Rico, 103
Cuomo, Andrew, 121, 137–38

decentralized production, 164
demand, 19, 109, 122, 186, 189, 195, 197; and duck curve, 189; energy mix, 193; in New York City, 53–57; peak, 20, 55, 70, 109
de Onís, Catalina M., 102, 103
Detroit, Michigan, 161–62; Covid-19 pandemic, 167–68; high-efficiency buildings, 162; metropolitan area, 167–68; power outages, 167–68. *See also* Highland Park, Michigan
Detroit Innovation Fellowship, 174
Detroit 2030 Project, 161
Dialogo. *See* Comite Dialogo Ambiental, Inc.
digital revolution, 165
Dirul, Ali, 162
disaster capitalism, 102–3
distributed generation, 17, 19, 35, 67, 77, 84, 87–88, 96, 108, 119, 155, 165, 183, 206, 207
diversity, 6–7
documentary film, 3
Dominican Republic, 112
Dovey, Rick, 24, 27–31
DTE Energy (DTE), 160, 167–68, 169, 170; annual revenue, 160; New York Stock Exchange, 160; rate policy, 175; reimbursements offer, 167; repossession of street lights, 170
duck curve, 19

Earthjustice, 97
earthquake and aftershocks, 121–22

The Eaton Blackout Tracker, 167
eco-village, 162, 163, 171–72, 177. *See also* Parker Village
Edison, Thomas, 67
El Coquí Community Center in Salinas, 98–99
Electric Utility Performance Report, 167
electricity, 13–20; bathtub metaphor, 13–15, 16, 193; grids and microgrids, 15–16; peer-to-peer marketplaces for, 165
electrocution from power lines, 41–43
emergency power: in Rockaways, 134–36; and solar kits, 115–18
energy audit, 116
energy burdens, 165–68, 173, 178
energy choices, 102
energy coloniality/colonialism, 101–2, 103, 118
energy democracy, 161, 164–65; academic research on, 164; compared with digital revolution, 165; concept, 161, 164; economic transformation and, 165; from energy poverty to, 165–68
energy inventory, 116
energy mix, 26, 57, 78, 83, 193
energy monopolies, 164–65
energy poverty, to energy democracy, 165–68
energy privilege, 102
energy production, 37
energy storage, 19, 195; GMP, 85–87; solar-powered water pumps, 156–57
energy transformation. *See* transformation of energy system
entrepreneur/entrepreneurship, 3, 81, 162; social, 81, 82
environmental communications, 101–2
environmental justice, 29, 52, 58–59, 61, 71, 72, 159, 161, 166, 168. *See also* Renewable Rikers; Soulardarity
equity, 6–7, 52, 59, 69, 81, 159, 165, 166
exodus, 103, 142

Famadou, Karanja, 162
FEMA, 105, 141, 153
Fiordaliso, Joseph, 44
Fishermen's Energy, 23, 24, 31–36
Ford, Henry, 171
Ford Model-T. *See* Model T of Ford
Ford Motor Company, 159, 168–69
fossil fuel, 5, 6, 7, 18, 32, 37, 38, 56, 85, 151, 161, 206; advantage, 194; climate change and, 130, 132; energy democracy approach, 165; Hurricane María and, 156–57; killing birds or impact on wildlife, 40–43; New York City, 57, 65, 67; pollution from, 6, 55, 95, 200
fossil fuel-burning power plants, 195
Fox, Jeanne, 24, 31–32, 45
FUD (Fear, Uncertainty, and Doubt) narrative, 37–41

Gallagher, Paul, 24, 33–35, 43, 45
gender gap, 7
geothermal power, 1–2, 7, 41, 51, 161, 165, 187, 197, 206
Gigantiello, Anthony "Tony," 53, 58, 59–60, 67
gigawatts (GW), 19
Glamour, 145–46
Goodman, Carolyn, 190–92
Goodman, Oscar, 190
Green Mountain Power (GMP), 77–91; as "America's most innovative utility," 77; as certified B Corporation utility, 81–82; customer service, 77–78; energy storage, 85–87; innovations, 82–85; internal cultural changes, 89–91; as largest electric distribution companies, 79–80; public reception and opposition, 88–89; redesigned solar and wind projects, 87–88; "smart meter" program, 80–81
Green New Deal, New York City, 62–66, 72, 207
grid attached system, 19, 99, 100

grid failures, 129–30
grid-tied, 99, 100–101

Higgins Generating Station, 194–95
Highland Park, Michigan, 159–79; as "City of Trees," 171; households below poverty line, 167; as poorest city, 171; population, 159, 168–69; solar ordinance, 176; streetlights, 169–79. *See also* Parker Village; solar-powered streetlights; Soulardarity
Hispanic Federation, 138
"How to Power a City" (documentary), 3
Hurricane Irene, 27
Hurricane Irma, 101, 132, 133, 140
Hurricane Katrina, 102, 133
Hurricane Maria, 87, 95–108, 113, 115, 206; and La Riviera, 149–57; Massol-Deyá on, 156–57; threat of disaster capitalism, 103
Hurricane Sandy, 27, 29, 33, 52, 130, 132, 134–35
hurricanes, 131–33; cold water, 133; rebuilding and disaster relief efforts, 133; warm water, 132–33
Hyde, Robert Wylie, 143

IDEBAJO (Initiative for the Eco-Development of the Jobos Bay), 97, 110, 117
Independent Commission on New York City Criminal Justice and Incarceration Reform, 69
Independent System Operator (ISO), 17
India, 151
Indian Point (nuclear power plant), New York, 57
information gathering, 4–5
installed capacity, 186
intermittency, 85
International Partnering Forum for Wind, 25
Irizarry-Rivera, Agustín, 108, 117

Isabela, Puerto Rico, 129–47. *See also* Solar Libre in Isabela
islanding off, 16, 19

Jobos Bay, Puerto Rico, 110
Jobos Bay National Estuarine Research Reserve, 110
Jones Act, 138–40

kilowatts (KW), 19
Kingdom Wind Farm, 89
Klein, Naomi, 102
knowledge, 118; community, 107, 112; as power, 177–79
knowledge gap, 7
Koeppel, Jackson, 162, 173

land use, 1, 106
La Riviera, Puerto Rico, 149–57
Las Dunas Guest House, 140
Las Palmas, 102
Las Vegas, Nevada, 188–95; demand and duck curve, 189; energy storage, 195; Higgins Generating Station, 194–95; LED lights, 193; Mayor Goodman, 186, 190–92; power purchase agreement, 193; retail load, 193; sustainability measures, 190–91; technology and collaboration, 192–95
leadership, 2
LED bulbs/lights, 164, 174, 191, 193
Lippman Report, 69–70
load, 19
Lowell, Vermont, 17
low-income families, 7

Mandalay Bay casino, 198–99
Mapp, Kenneth, 87, 133, 137
Massol-Deyá, Arturo, 156–57
MAVI Arecibo, 144
mega-region, 46
megawatts (MW), 19
Meyer, Thomas, 131, 136, 138, 140–42

Meyer, Walter, 131, 134, 135, 136, 137–42, 146, 150
Michigan: Covid-19 outbreaks, 167–68; power outages, 167–68
Michigan Public Service Commission and Legislature, 175–76
microgrids, 15–16, 19, 27, 29, 30, 84, 88, 101, 104, 107, 119, 155, 156
MIT, 117
mixed generation, 17
Moapa Band of Paiutes, 197
Moapa Southern Paiute Solar Project, 197
Model T of Ford, 159, 169
Mujeres Ayudando Madres, 145–46
Murphy, Gregg, 150
Murphy, Phil, 43

Nandi's Knowledge Café, 162, 173
narrative inquiry method, 4
National Audubon Society, 37–38
National Resources Defense Council, 61
National Wildlife Federation, 41
natural gas, 57, 120, 195
Nellis Air Force Base, 198
Nellis Solar Array II Generating Station, 198
net metering, 16–17, 19, 29, 119, 121, 196, 199
net zero, 16
Nevada, 183–201; casinos, 198–99; citizen-led campaign, 196; geothermal utility power, 197; nuclear dumping, 199–201; "renewables first" policy, 194; residential net metering, 196; tribal lands and solar, 197–98
New Energy Consultants, 154
New Jersey, 23, 27; clean energy in, 31; Hurricane Sandy, 27, 29, 33; offshore wind industry, 24; Regional Greenhouse Gas Initiative, 35; solar power, 31–32. *See also* Atlantic City, NJ
New Jersey Board of Public Utilities, 23, 25

New Jersey Renews, 43
New York City, 51–72; buildings, 53–54; coal-burning plants, 67; coronavirus impacting electrical use, 56–57; Costa (City Council member), 53, 59, 61–66, 67, 70–72; Green New Deal, 62–66, 72, 207; Indian Point shutting down, 57; natural gas plants, 57; population, 55; power hungry demand, 53–57; power plants, 55; Renewable Rikers, 69–72; solar power systems in, 11, 12; sustainability networks in, 11
New York City Climate Law Tracker, 65
New York State: clean energy initiatives, 54–55; offshore wind, 54–55, 57; peaker plants, 59
New York State Independent System Operator (NYISO), 56
New York State Reliability Council (NYSRC), 56
New York Stock Exchange, 160
Nichols, Shimekia, 162, 172–73, 175, 176
North American Board of Certified Energy Practitioners (NABCEP), 142
nuclear dumping in Nevada, 199–201
nuclear power, 1–2, 16, 83, 112, 164, 199–201; as clean energy, 1–2, 200; Indian Point, 57; killing birds, 40, 42; pollution, 6
NV Energy, 79, 96, 185, 193, 194–98, 200, 201

Ocasio-Cortez, Alexandria, 71
off-grid, defined, 20
off-grid solar power, 177. *See also* solar power systems
off-grid water pumps. *See* solar-powered water pumps
offshore wind farms, 23–46; birds and turbines, 36–40; economic benefits, 44–45; onshore *vs.*, 39–40
O'Neill-Carrillo, Efrain, 97, 108

onshore wind farms, 17, 26, 28, 39–40, 45; offshore *vs.*, 39–40
output, 55, 194; solar panel, 20, 28; turbines, 28

Pacific states: offshore wind development, 23–24
Palmas Del Mar, 100
Paris climate accord, 162
Parker Village, 162, 163–64, 171–72, 174; as an eco-village, 177; a food distribution center for Covid-19 relief, 178; solar-powered lighting, 177–79
peak times/peak demand, 20, 55, 70, 109. *See also* demand
Pearl Street Station, 67
peer-to-peer marketplace, 165
peer-to-peer marketplaces for electricity, 165
Pew Research Center, 103
Polar Bear Sustainable Energy, 177
Poletti Power Plant, Astoria, Queens, 58, 59–61, 65
pollution, 130
population loss, 169, 189
Powell, Mary, 78, 79, 80, 81, 82–89, 91
power island, 16
power outage, 130, 167–68
power purchase agreements, 17, 20, 198
privilege, 102. *See also* energy privilege
prosumer, 117–18
public-private partnerships, 192
Puerto Rico: Act 17, 119–20; electricity use, 109; energy consumption per capita, 109; new energy law and new challenges, 119–22; Queremos Sol, 119; solar power need, 108–9; toxic waste and pollution, 109–15
Puerto Rico Electric Power Authority (PREPA), 95; federal emergency funds, 122; Integrated Resource Plan (IRP), 121
Puerto Rico Energy Bureau, 117
Puerto Rico Public Service Regulatory Board, 117

Queremos Sol, 100, 102, 106, 119, 120, 121

Ravenswood Power Plant, 59
Recorder Riker, 58
Regional Greenhouse Gas Initiative, 35
Regional Transmission Organization (RTO), 17
Reid Gardner Generating Station, 197
renewable energy. *See* clean and renewable energy
Renewable Rikers, 53, 57, 59, 69–72
"renewables first" policy in, 194
repossession, 169–71
residential batteries, 58–59, 156
residential solar system, 7, 11–12, 16–17, 99–100; net metering, 16–17, 19, 29, 119, 121, 196, 199; in Nevada, 185, 196
Rhode Island, 23; offshore wind farm, 23
Rikers Island, 58–59, 69–72; Dutch government, 58; tribal societies, 58; using as a jail, 58
Rockaways, emergency power in, 134–36
Rocky Mountain Institute, 120
rooftop: solar, 9, 13, 17, 27, 53, 54, 56, 68, 84, 95–100, 105–9; wind turbines, 12, 54, 55
Rosselló, Ricardo, 119, 139–40
Ryter Cooperative Industries (RCI), 162, 172, 173–74

Sabin Center for Climate Change Law, 55, 65
Salinas, Puerto Rico, 95–122. *See also* Puerto Rico
Santana, Juan, 151, 152, 155
Santiago, Ruth, 96–97, 99, 100–101, 105–18, 120
Santos, Nelson, 110
Shannon, Juan, 162, 172, 174, 177, 178
Siemens, 120
social entrepreneurship, 81, 82

solar arrays, 101
solar companies, 103
solar donations, 104
Solar Energy Industries Association (SEIA), 104, 186
solar farms, 12
Solar Libre in Isabela, 99, 133–42; clean energy systems, 134; design, 145–46; ground operations, 140–42; Jones Act, 138–40; mobile solar generator, 142; training program, 134, 142–46
solar panel output, 20, 28
solar panels, 16
solar power systems, 11–12, 99–101; emergency, 115–18; empowerment, 101–4; off-grid, 99–100; residential rooftop, 99; size of, 99
solar-powered streetlights, 171–79; defined, 163–64; statewide coalition, 175–77
solar-powered water pumps, 150, 151–56; energy storage systems, 156–57; larger scale impact, 155–56
solar-powered Wifi, 176
solar PV (photovoltaic), 20
solar tax, 121–22
solar thermal, 20
Solutions Journalism (SJ), 5–6; tenets of, 6
Solutions Journalism Network, 6
Soulardarity, 159, 160, 161–62; "Ask Me About Power" slogan, 171; community campaigns, 171; education programs, 162; solar-powered streetlights, 171, 172–75; streetlight program, 162; Wifi-enabled solar-powered streetlights, 176; Work for Me DTE! campaign, 175–76
Staten Island, 16–17
streetlights: repossession, 169–71; solar-powered, 163–64, 171–79

Technological Institute of Guayama, 116

terminologies, 18–20
Territory Profile and Energy Estimate whitepaper, 109
Tesla, 104, 115; Powerwall batteries, 77, 84, 85
Thomas, Roberto, 97
Time for Turbines, 24, 43–44
toxic waste and pollution, 109–15; advocacy campaigns, 110; coal ash, 110–15
training program of Solar Libre, 134, 142–46; apprentices in, 143–44; classroom education, 143; solar panels at MAVI, 144
transformation of energy system, 2, 5, 131, 161, 162, 164–65, 186, 201, 205–7; benefits from, 206–7; Coquí Solar, 98–109; Green Mountain Power (GMP), 77–91

United Nations: sustainable development goals, 151
University of Puerto Rico, Mayagüez, 97
U.S. Air Force, 198
U.S. Army Corps of Engineers, 141
U.S. Census, 103
U.S. Conference of Mayors, 191
U.S. Energy Information Administration, 109
U.S. Fish and Wildlife Service's National Forensics Lab, 41

Vermont, 45; climate and energy policy, 80; Green Mountain Power (GMP), 77–91; sun for solar, 88; utilities in, 77
Vermont Law School, 80
Vieques, Puerto Rico, 103
Virginia: offshore wind farm, 23

Water Mission, 149–51; battery-and-solar systems, 150; solar-powered water pumps, 150, 153–56

water pumps, 23, 149–55; diesel-burning, 151; fossil fuel, 151; solar-powered, 150, 151–57
Whitefish, 103
Wifi-enabled solar-powered streetlights, 176
wildlife: fossil fuel production and imapct on, 40–41. *See also* birds, death of
wind energy/farms/power, 12; offshore, 23–47; onshore, 17, 26, 28, 39–40, 45
wind turbines, 12. *See also* birds, death of
Work for Me DTE! campaign, 175–76

Yarotek, 100, 101
YouTube, 170
Yucca Mountain nuclear waste repository, 199–201

zero water discharge, 111

About the Author

Melanie J. La Rosa is a filmmaker and professor of media production at Pace University. She has made numerous documentaries, most recent being, "How to Power a City." She has received numerous grants for filmmaking and education, from the New York State Council on the Arts, the Periclean/Andrew W. Mellon Faculty Leadership Award, the Queens Council on the Arts, the Brooklyn Arts Council, and the Solutions Journalism Network. She attended Temple University and the University of Michigan and grew up near Lansing, MI. She has lived in the Bay Area, Philadelphia, Washington DC, and now resides in Queens, New York City.

www.ingramcontent.com/pod-product-compliance
Lightning Source LLC
Chambersburg PA
CBHW061711300426
44115CB00014B/2637